The Use of Recovery Factors
in Trace Analysis

The Use of Recovery Factors in Trace Analysis

Edited by

M. Parkany
International Organization for Standardization,
Geneva

THE ROYAL
SOCIETY OF
CHEMISTRY
Information
Services

The proceedings of the Seventh International Harmonization Symposium on a Protocol for Recovery Factors held in Orlando, Florida on 4–5 September 1996.

Special Publication No. 184

ISBN 0-85404-736-0

A catalogue record for this book is available from the British Library.

Published by The Royal Society of Chemistry,
Thomas Graham House, Science Park, Milton Road,
Cambridge CB4 4WF, UK

Printed by Bookcraft (Bath) Ltd

Foreword

by Dr. Lawrence D. Eicher, Secretary-General, ISO
(International Organization for Standardization)

This Seventh International Harmonization Symposium on the use of recovery factors is expected to examine all facets of the problem from the scientific, practical and legal points of view. Most of the lectures rightly focus on agricultural and food chemistry, since this is a problematic issue and the Codex Alimentarius Commission (CAC) expects a viable protocol (based on consensus) that they can adopt.

ISO is well aware that this problem has a detrimental effect on global trade in agricultural and food products. At this time there are no accepted standards or guidelines on recovery factors that might be followed. Furthermore, it is most unlikely that any single ruling could be applied in all cases. The practical examples mentioned by the lecturers will certainly help establish a list of problems together with analysis grouped according to the recommended or deprecated use of recovery factors.

In a certain number of lectures, cases are quoted where regulatory and/or court decisions depend on the application (or non-application) of recovery factors and their enormous economic consequences.

We standardizers know very well that soon there will be a demand to have clauses in the relevant international test method standards give unambiguous descriptions of recovery factors, and we are prepared for it.

There is no doubt that other analytical methods will have to be considered for the same problem. As an example, I mention human toxicology and specifically the field of Occupational Health and Safety (OHS) that I feel will soon be coming into the limelight.

It may be, however, that the most urgent demand will stem from the requirements of the first of the ISO 14000 series of environmental management standards on EMS (Environmental Management Systems) that will be published by the time of the Symposium. These standards, from the point of view of this Symposium, require valid and exact measurement methods, and clear indications on the use of recovery factors in order to establish traceable, documented facts able to stand up to any scientific scrutiny in the fields of

    a) emissions to air;
    b) releases to water;
    c) waste management;
    d) contamination of land;
    e) impact on communities;
    f) use of raw materials and natural resources;
    g) other environmental issues.

Furthermore, following the EMS standards on the Continual Improvement Spiral, the Checking and Corrective Action Stage requires (among other actions) Monitoring and Measurement. No one will argue that valid, indisputable procedures have to be applied if the standard is to be implemented. Pushing this logic one step further, all will also agree that the correct use of recovery factors, as for example in the case of the analysis of contaminated land, has to be established for an implementation of the ISO 14000 series. How can auditors/assessors establish the implementation of these standards if correct measurements and calculations are not standardized? How can a company be certified to these standards if the cornerstone of *measurement* is not laid down? I foresee strong pressures from the market to have such certification, and this in turn will create a demand for the Working Party on Harmonization to finalize the protocol on recovery factors as soon as possible.

Introduction

by Dr. Michael Parkany, Consultant, ISO
(International Organization for Standardization)

This is the seventh in the series of International Symposia for Harmonization in the field of Quality Assurance in Analytical Chemistry.

This Seventh Symposium (planned for 4-5 September 1996 in Orlando, Florida, USA) is devoted entirely to the subject: When is it justified to use recovery factors? It is evident that in this type of analysis the description of the analytical method must be clear on this and must give precise instructions. When preparing these Proceedings it was found that at present standards and protocols on analytical procedures very rarely fulfill this requirement. This is why it is so problematic to obtain a real picture of whether and when the use of recovery factors is forbidden, tolerated, encouraged or required. In fact, very often, the choice is left for the analyst in charge. Evidently judges in courts cannot accept such a situation, since they cannot base their judgements on conflicting interpretation of standards, neither are they in a position to improve the texts of standards and protocols for analytical procedures. They rightly require that these documents give clear instructions.

The lectures in these Proceedings review the present situation and offer practical solutions based on the experience of analysts of the field. Furthermore, a questionnaire was circulated to about eight hundred laboratories in order to establish present day practice. On the basis of answers to this questionnaire and of lectures at the Symposium, a Draft Protocol will be prepared by the IUPAC Interdivisional Working Party on Harmonization of Quality Assurance Schemes for Analytical Laboratories. This Protocol, when accepted, will be adopted by the Codex Alimentarius Commission (FAO -WHO). Finally a majority of countries will incorporate it into their legal systems.

This is why it is extremely important to be very careful when preparing a Protocol that will influence analytical results and important decisions in the field of food, agriculture, environment, toxicology, etc.

Standards writers should be involved and they should participate in the work, or at least be informed. Since the sponsors of the Symposium are ISO, IUPAC and AOAC International, the main international partners are involved.

I am very pleased that among the illustrious scientists involved we have the honour to have a lecture offered by Dr. William Horwitz, who has the longest career among us in this field and whom we regard as our Master/teacher. His experience is a rich treasure that he generously makes available to all of us.

I wish to acknowledge the help of the active members of the IUPAC Interdivisional Working Party, among them first of all Dr. Roger Wood and Dr. Michael Thompson who accepted the task of the preparation of the questionnaire, the evaluation of the answers and the drafting of the Protocol.

We all extend our thanks to the authors for providing their manuscripts,and to Mrs. J. C. Freshwater of the Royal Society of Chemistry for arranging this special publication.

Contents

The Use and Misuse of Recovery Factors in Analytical Chemistry
William Horwitz   1

Considerations in the Estimation of Recovery in Inorganic Analysis   5
Robert W. Dabeka and Milan Ihnat

Use of Radioisotopic Tracers for Determination of Recovery Factors   24
(Chemical Yields) in Trace Element Determinations by Radiochemical
Neutron Activation Analysis (RNAA)
A. Fajgelj and A.R. Byrne

Measurement Uncertainty: the Key to the Use of Recovery Factors?   30
Stephen L.R. Ellison and Alex Williams

Varying Uses of Recovery Factors in US EPA FIFRA Registration Studies   38
Del A. Koch, Patrick A. Noland and Loren C. Schrier

Use of Recovery Data in Veterinary Drug Residue Analysis   42
James D. MacNeil, Valerie K. Martz and Joe O.K. Boison

Development of Recovery Factors for Analysis of Organochlorine Pesticides   48
in Water Samples by Using a Micro-Extraction Technique
R. Boonyatumanond, M.S. Tabucanon, P. Prinyatanakun and A. Jaksakul

Analytical Recoveries and Their Use for Correction   59
J.F. Kay

Report of the 20th Session of the Codex Committee on Methods of Analysis   63
and Sampling

Selective List of Relevant International Standards   101

Addresses of Authors   125

Subject Index   127

The Use and Misuse of Recovery Factors in Analytical Chemistry

William Horwitz

US FOOD AND DRUG ADMINISTRATION, HFS-500, 200 C STREET SW, WASHINGTON, DC 20204, USA

The use of a correction to a measured value (recovery factor) is appropriate only if the deviation is the result of a systematic error and the purpose is to compensate for a constant error in operations or methodology (bias). If both the measured value and the correction factor are subject to random error, the standard deviation of their combination is the square root of the sum of the squares of the two standard deviations, from the law of propagation of error. In such cases it is not appropriate to use a correction factor.

1. INTRODUCTION

At first glance, the use of recovery factors in analytical chemistry is attractive. Thinking in terms of the solubility of a precipitate, the obvious unleachable color left on a filter paper, or the undesirable presence of inorganic salts in organic dietary fiber, the immediate reaction is to "correct" for such unwanted effects. In fact the definition of "result of a measurement" in the "International Vocabulary of Basic and General terms in Metrology"[1] as the "Value attributed to a measurand, obtained by measurement," states in a note that, "When a result is given, it should be made clear whether it refers to: the indication, the *uncorrected* result, the *corrected* result and whether several values are averaged" [emphasis added]. It continues with definitions for "uncorrected result" and for "corrected result," as the "the result of a measurement before [after] correction *for systematic error*" [emphasis added]. Therefore, the use of a correction to a measured value is appropriate only if the deviation is the result of a systematic error.

2. EXAMPLE

Let us take a simple example. Consider that you make a determination of a pesticide residue, XCl, in a test sample and find 1.0 mg/kg. You also take an identical material not containing the analyte [the blank (control)], add 1.0 mg XCl/kg, perform the analysis, and obtain 0.9 mg/kg for this spiked material. Do you conclude that your recovery is 90%?

Table 1 *Example of Recovery Calculations from Two Determinations (No. 1 and 2)*
of a Residue of XCl and Two Determinations (No. I and II) of 1.0 mg/kg XCl
Added to a Blank Matrix:

Analytical Results, mg/kg, 1 = 1.0   I = 0.9
                                             2 = 0.9   II = 0.8
                                          ave. = 0.95 AVE. = 0.85

Combinations				Recovery,
Control	Minus	Test	= Loss	% loss
I(0.9)	-	1(1.0)	0.1	10
I(0.9)	-	2(0.9)	0.0	0
I(0.9)	- ave.(0.95)		0.05	5
II(0.8)	-	1(1.0)	0.2	20
II(0.8)	-	2(0.9)	0.1	10
II(0.8)	- ave.(0.8)		0.95	15
AVE.(0.85) -		1(1.0)	0.15	15
AVE.(0.85) -		2(0.9)	0.05	5
AVE.(0.85) - ave.(0.85)			0.95	10

The answer is: If you stop here, yes. But if you repeat either or both analyses, you immediately discover a calculation problem. If the second result on the test sample is 0.9 mg XCl/kg, what number do you use as the basis for your calculation? Further, if a second control determination gives 0.8, which of the possible combinations do you use? Table 1 gives the possibilities in excruciating detail, together with the additional possibility of using the averages.

Most researchers will agree that the last calculation, using the averages, gives the "best" estimate of a 10% loss or a 90% recovery *in the long run*. But with single determinations of the tests and controls, any of the other values are equally likely. Consequently the analyst could report recoveries for this determination of anything from 80 to 100%. If values 1 and I were reported by analyst A and values 2 and II were reported by analyst B, they would both probably blame the test samples as the cause of the discrepancies. Yet all of these values can arise from sampling normal distributions of analytical values with means of 0.95 and 0.85, respectively, and standard deviations of about 0.15. The answer then comes down to this: You can correct for bias in the long run, but you have no control over individual value corrections.

3. PROPAGATION OF ERROR CONSIDERATIONS

This example provides a simplified illustration of the law of propagation of error. If both values entering into a final result are subject to random error, the standard deviation of their combination is the square root of the sum of the squares of the two standard deviations. Further, if the input values are averages the standard deviation of the average must be divided by the square root of the number of values in the average. Therefore, the standard deviation of the combination will include this additional factor.

For the example given, and assuming a standard deviation of 10%,

$$s_{AB} = \sqrt{(s_A^2 + s_B^2)} = \sqrt{(10^2 + 10^2)} = 14(\%)$$

If each value is the average of 2 values, the standard deviation of the result is sAB divided by $\sqrt{2}$ = 10%. The effect of the law of propagation of error is offset here only by doubling the amount of work performed.

4. APPROPRIATE AND INAPPROPRIATE USE OF RECOVERY FACTORS

The use of recovery factors is appropriate only when the purpose is to compensate for a constant error in operations or methodology (bias). This situation is most frequently encountered in extractions, both solid and liquid phase. Although it would appear that the recovery is constant, i.e., the initial isolation-of-analyte step of an analytical method fails to extract a constant amount, the recovery test concentration is usually measured by a procedure that is just as variable as the determination of the analyte concentration as an unknown, although the quantity added is fixed. Application of a constant recovery factor under such circumstances may reduce bias (difference from the true value in the long run) at the expense of precision. When a corrected result is reported, the variability of the measurement of the recovery is propagated into the variability of the determination, resulting in a relative standard deviation (RSD) of 1.4 ($\sqrt{2}$) times the RSD of the uncorrected individual value. When an internal standard is used, the same considerations still apply. If the internal standard has been conducted through the entire procedure, as it should be, and if it is measured in the same way as the test analyte, it is subject to the same sources of random error.

In pesticide residue analysis, the average long-term recovery at the ppm (10^{-6}) level is 85-90%. However, a recovery factor is typically not applied for several reasons:

(1) The recovery of controls conducted simultaneously is just as variable as the individual determinations, e.g., 70-110%. Reports and publications typically present both sets of data independently for use by reviewers as they wish.

(2) Regulatory specifications usually do not need "true values." The tolerances that have been established already have taken the recovery into consideration. They are promulgated after a review of data supplied by the petitioner (at least for pesticide and veterinary drug residues in the US) that have been obtained from plants and tissues by the same methods that will be used in enforcement. Therefore the specification has already compensated for the recovery. The net result of the use of a correction factor in these cases is to provide higher values for the residues than were intended.

This [preadjustment] is most easily seen in the use of empirical methods (CODEX Type I methods) in establishing food composition specifications. When a composition specification is established (e.g., moisture, fat) the method is usually specified so that the results obtained for enforcement will correspond to the values that have been established as the required limits *by that method.*

(3) Those methods that use internal standards have an apparent "built-in" correction factor. However, this correction may be deceptive when the internal standard is not carried through the entire procedure, especially when the internal standard is added after the native residue has been extracted.

One other term is important in the calculation of recovery factors--the value to be used in the denominator, when a multiplicative constant is applied. Recovery can be calculated on the basis of (1) only the added analyte, with the native analyte in the matrix ignored (designated as marginal recovery), and (2) the total analyte present, both native and added (designated as total recovery). When the quantity of analyte in the matrix (blank or control) is essentially 0, both calculations give the same value; when the native amount present [the incurred residue] is significant, the "method of additions" becomes appropriate to determine the quantity originally present in the matrix, so that the recovery effort is more than a mere additional measurement. In such a case, more quantities must be measured, resulting in additional variability from propagation of error and thereby diluting the bias-correcting effort. Thus the operation may become a self-defeating exercise.

There is a further complicating effect when biological systems are involved. In such a case the administrator is not interested in minor analytical niceties. The decision-maker is interested in recoveries of the parent compound in cases of feeding veterinary drugs to animals or applying pesticides to crops, whereas the analyst is interested in the recovery of the particular metabolite that is easiest to measure. Therefore the laboratory manager must be very sensitive to the nature of the recovery being reported. Do the recoveries reported properly reflect the compound of interest, or something far down the reaction chain?

There are several other technical problems with the concept of correction factors, such as the mathematics of a multiplicative versus an additive factor, and the statistical problem of a standard deviation that varies with the concentration added, which we will merely mention, but not discuss.

Therefore, what appears on the surface as a rather trivial exercise can have considerable ramifications. By no means does this discussion advise that recovery of added analyte determinations not be performed. On the contrary, recovery studies are an essential component of quality assurance systems. Rather the discussion suggests reporting recovery results separately to permit an administrator to correct or not, whichever is appropriate for the purpose.

5. REFERENCES

'International Vocabulary of Basic and General Terms in Metrology', Second Edition, International Organization for Standardization, Geneva, 1993.

Considerations in the Estimation of Recovery in Inorganic Analysis

Robert W. Dabeka[1] and Milan Ihnat[2]

[1]FOOD RESEARCH DIVISION, HEALTH PROTECTION BRANCH, HEALTH CANADA, OTTAWA, ONTARIO K1A 0L2, CANADA

[2]CENTRE FOR LAND AND BIOLOGICAL RESOURCES RESEARCH, RESEARCH BRANCH, AGRICULTURE AND AGRI-FOOD CANADA, OTTAWA, ONTARIO K1A 0C6, CANADA

1. INTRODUCTION

Harmonized, reliable analytical measurements are vital in routine analysis, research, comparison of methods and information, standardization of laboratory performances and legal compliance with government regulations for national requirements and international trade. Inclusion of appropriate tests of analyte recovery into the analytical scheme, incorporating good methods and other aspects of a quality control and quality assurance program, is integral to the ability to properly assess, monitor and maintain good analytical data quality.

The determination and use of recovery factors is currently a controversial issue among analytical chemists. Various concepts are held and approaches directed, in organic and inorganic determinations, to the estimation and use of recovery factors. A great many parameters impinge on the philosophy and practice of the determination and use of recovery factors which must be considered and evaluated in an effort to arrive at a consensus of a harmonized best practice.

The aim of this paper is to deal with some of the important parameters and issues related to recovery of inorganic elemental analytes, and to put forth some ideas, philosophy and approaches to stimulate and provoke thought and discussion by attendees at the Symposium in their quest for a consensus for harmonization. It will focus on general considerations for the use of recovery materials, estimation of recovery based on added and endogenous analytes, calculation of recovery factors and associated uncertainties, the impact of method and laboratory bias/systematic error on recovery and the application of recovery factors. Although the ideas herein relate to recovery considerations in general, the emphasis is on inorganic analysis, determination of elemental content at macro and trace levels. Furthermore, while extractable or bioavailable elemental concentrations are also of interest to some analysts, only total elemental concentrations are addressed.

2. GENERAL CONSIDERATIONS FOR USE OF RECOVERY MATERIALS

2.1 Sources of Error in Analytical Methods

The general lack of agreement among analytical results from different analysts and laboratories arises from numerous factors influencing the validity and reliability of the final numerical results. These factors can be broadly categorized as presampling, sampling, sample manipulation and measurement. Other important considerations such as contamination control, data quality control and the analyst transcend the above boundaries.

Presampling factors,[1] as far as biological specimens are concerned, include genetic predisposition, long term physiological influences such as age, sex, geographical and environmental factors, diet, pregnancy and lactation, short term physiological influences including circadian rhythms, recent meals, posture and stress, seasonal changes including physiologic and climatic influences, postmortem changes including cell swelling, imbibition and autolysis, intrinsic errors involving medication, haemolysis, subclinical conditions and medical restrictions in sampling, and correct choice of target organ.

Sampling refers to the entire collection of steps and considerations in procuring the sample such as identification of the population, establishment of a sampling model and plan, meaningful, proper and representative sampling using appropriate collection techniques, storage, transportation, and reduction of the gross sample to laboratory sample.[2-4]

Sample manipulation encompasses material and solution storage, decomposition, extraction and separation of analyte, volumetric ware verification, calibration, technique of use, sample drying and/or moisture determination and dilution schemes. Examples of sources of errors are: analyte volatilization losses in dry ashing or wet decomposition, incomplete destruction of matrix, recovery and analysis of insoluble residue, alteration of oxidation state during decomposition and extraction, contamination from ashing aids, acids and reagents, contamination from and losses to decomposition vessels, and incomplete separation/extraction. Measurement refers to the steps quantifying the amount of analyte present and covers the actual technique of signal determination, calibration, matrix effect management, data handling and calculation. It deals with selection of proper analytical techniques, instrument optimization, optimization of determinative technique performance characteristics and utilization procedures, correction for physical, chemical and background interferences, selection of calibrants (starting material purity and composition, preparation techniques of stock and working calibrants, verification, dilution schemes), selection of calibration solutions (single analyte, composite, matrix matching) and calibration technique (calibration curve, bracketing for high precision). The penultimate step in the scheme of analysis, preceding data interpretation and final use, is data handling and calculation which entails data and information recording and calculation, calibration curve fitting and calculation techniques, interpretation and evaluation (controls, statistical treatment, data presentation).

It is no wonder that such an extensive collection of potential pitfalls seriously impacts on data quality and typically imparts to it substantial questions of validity. Tests with recovery materials can monitor and control, to a good extent, the performance of the collection of laboratory procedures subsequent to the point of introduction of the material. Errors arising from activities occurring prior to this point of introduction, such as sampling, preservation, storage and presampling considerations are generally impossible to monitor by use of recovery tests.

2.2 Procedures for Use of Recovery Materials

The recovery material, which can be either a spiked sample or a native analyte - natural matrix Reference Material, can be incorporated into the analytical sequence in one of several different ways. The suggested, preferred mode is incorporation at random in the sequence of samples in the batch analyzed. In the case of spiked samples, those to be fortified and to be tested for recovery can be selected randomly using random numbers, spiked and again randomly included in the sequence. Alternatively, for logistic and convenience considerations, selected samples and their spikes can be physically adjacent to each other in the analytical sequence, or the spiked samples may be placed at either the beginning or end of the batch.

Selection of recovery level for monitoring recovery should consider the expected concentration of the analyte determined and the possible effect of concentration on recovery. As a first choice, the level should be at or near the enforcement limit or near the level actually present in the sample for most appropriate monitoring of method performance, particularly when dealing with highly homogeneous materials. Alternatively, the spike level can be a certain multiple of the limit of quantitation, limit of determination or limit of reporting. Recovery testing should be carried out at higher levels than the preceding limits if higher concentrations are anticipated in the material or there is a question of material inhomogeneity. If labour is a consideration (and it likely is under today's global fiscal constraints) single, rather than duplicate, spikes of different samples/materials are preferred instead of duplicate spikes of the same sample or material; additional effort should be devoted to covering a wider range of material undergoing analysis. Consideration of determination of recovery for each matrix type encountered in analysis or only on selected matrices, representative of the materials analyzed, depends on the criticality of the analyses.

2.3 Multielement Determinations

In many analyses nowadays, methods are geared toward multianalyte determinations for cost-effectiveness and in response to the requirements of clients. The determination of a recovery factor for one analyte, and its application to monitoring/correction of other analytes, assuming it to be constant and applicable to all analytes determined, is unsuitable and strongly discouraged. This is equally true whether dealing with different or related organic analytes or elements. Although there will be a semblance of similarity in chemical behaviour among related analytes throughout the various chemical reactions constituting the method, behavioral differences can be significant. Error types and magnitudes can be quite specific to each analyte. The various errors in sampling, sample manipulation and measurement impact differently on recovery of analytes from the sample bulk matrix. It is to be expected that collection techniques, sample storage and transportation, reduction of the gross sample to laboratory sample, sample manipulation, analyte volatilization during decomposition, incomplete extraction/separation, analyte retention by solid residue from incomplete destruction of matrix, alteration of oxidation state during decomposition and extraction, contamination, calibration, matrix effect management, selection of proper analytical technique, correction for physical, chemical and background interferences, starting calibrant material purity and composition and specific calculation details can all impinge differently on each analyte of interest. Thus when multielement determinations are conducted it is vital to determine an individual recovery factor for each analyte.

2.4 Preliminary Requirements

In order to produce valid analytical data and to properly and cost- effectively make use of recovery, it is essential that compliance with several prerequisites be established with the principal ones being correct analytical method and quality control.

An appropriate analytical method must be applied to the task on hand, by appropriately qualified and trained personnel in a suitable physical and administrative environment. Suitable physical environment refers to the equipment, materials, reagents and laboratory conditions necessary for the proper execution of the method; suitable administrative environment includes understanding of and support for appropriate data quality by the analyst's supervisor and all other managers. The role of the analyst is of direct paramount importance; good analysis and good analyst go hand in hand. Analyst training, experience, familiarity with the problem on hand, skill, attitude, motivation and judgement are necessary for satisfactory solution of analytical problems.

Suitable quality control/quality assurance procedures should be routinely in use and the need for appropriately reliable analytical information must be recognized. The analytical system must be in a state of statistical control; ie operating optimally and consistently generating acceptable data.

When dealing with the determination of total concentrations of elements, that is, the sum of all the element concentrations in all material (sample) phases and molecular species, it must be ascertained that the method is in fact measuring all of the element. The sample decomposition procedure must bring into solution all of the material with no grains or insoluble fraction left behind (eg. Ihnat [5]). In addition, the element must be in the correct oxidation state required by the various chemical reactions constituting the procedure.

3. DETERMINATION OF RECOVERY BASED ON ADDED ANALYTE

3.1 Applicability of the Spiking Approach

Frequently determination of recovery is based on the addition of the analyte being sought to a sample of material being analyzed. In this approach, a known quantity of pure analyte is introduced at some stage in the analysis process, the sample/analyte combination is carried through the analysis and comparison of results with the baseline value determined for the sample gives an estimate of recovery. The nature of the added analyte, selected from available elements or compounds used for calibrant preparation, is not necessarily identical to or representative of the nature and form of the native analyte occurring in the natural material being analyzed. This consideration is true for both organic and inorganic constituents. Thus, in principle, recovery estimated in this manner is not strictly accurate and should be regarded as solely an estimate. With organic analytes, reliance on spiking with analyte(s) of interest is, at times, the sole alternative for recovery determination.

Dependence of measured recovery on the nature of the element is not expected to be significant when sample destruction is complete, the material has been quantitatively brought into solution and all of the element is in solution and available for reaction and detection as required. It is also not expected to matter for certain instances of incomplete decomposition when the characteristics of undissolved residues are known. For example, if it is acceptable that Pb retained by silicate or siliceous residue of plant foodstuffs is not recovered, detected and measured by the applicable test method, then recovery of added Pb will be a reasonable measure of recovery. However, generally, the nature of the element should be taken as an important consideration when incomplete digestion is encountered. In instances of incomplete destruction of matrix, the time from spiking to analysis may also be an important factor.

3.2 Procedures for Recovery Determinations Based on Added Analyte

Consideration of determination of recovery for each matrix type encountered in analysis or only on selected matrices representative of the materials scheduled for analysis depends on the criticality of the analyses. A preliminary semiquantitative analysis of selected samples would be advantageous to establish at least an estimate of matrix composition to facilitate selection of the typical, representative and most suitable materials for spiking from among the lot to be analyzed. This usually may not be feasible unless the analyst has access to high throughput multielement analytical techniques and the analyst must decide whether it is worthwhile to devote such additional effort to cover a wider range of material.

Introduction of analyte can be *via* a given volume of a solution of chosen concentration or with a known mass of dry analyte or compound. Solution addition is substantially a more feasible and convenient technique on account of the generally very small quantities of analyte required. Manipulative constraints in weighing and dealing with minute or microscopic amounts of solid analyte or compounds thereof, preclude sufficient accuracy with this approach.

The nature of the recovery solution(s) used for spiking is also dependent on the criticality of the analysis and the analyst's judgement and convenience. As a first choice, the recovery solution should be prepared, independently of the calibrants, from the pure solid element/compound weighed from the supplier's container, or from the concentrated stock solution from a reputable government agency or commercial supplier. This approach will ensure a more independent and accurate determination of recovery. A second choice is separate preparation of the recovery solution, at the required concentration, from the same single element or composite stock solution used for preparation of calibrants. Yet a third suggestion is simply use of a solution identical to one of those used for calibration, ie. one of the higher concentration calibration solutions. For multielement analyses, it would be efficient and cost-effective to use one aliquot of a suitable multielement recovery solution containing the multiple elements at appropriate concentrations. Addition of spike in the analytical scheme should be at the earliest opportunity.

Spiking levels should be selected to represent the expected level of analyte to be measured or a certain multiple of the enforcement / quantitation / determination / reporting limit. To incur acceptable error in the recovery factor, the ratio of added analyte to analyte present in the material should be several-fold the concentration of naturally-occurring analyte with the actual ratio depending on circumstances. On the other hand, from the point of view of possible differences in analyte/matrix interactions at different, non-natural ratios, spiking levels should be closer to native content. Due consideration should be given to ensuring that the response of the spiked sample falls on the calibration scale to permit adherence to identical conditions of dilution, calibration etc. as for actual sample. How much one can one deviate from this again depends on the specific circumstances; however, the greater the ratio the greater the certainty of recovery factor measurement.

It may be deemed important to carry out a sufficient number of repeat measurements at each concentration level in order to get a good estimate of the uncertainty, a parameter we believe essential in estimating recovery. Repeats are especially necessary when ratios of spike to native or total level are small or unfavourable due to existing native levels leading to high uncertainty in the recovery factors. Assessment of amount of analyte (native level) already present in the sample prior to conduction of recovery tests is absolutely vital as proper definition of spike recovery refers to recovery of the added quantity. The rate of incorporation of spiked samples is at the discretion of the analyst and could range from less than one spiked sample per 100 samples (1/100) to more than 1/10 depending on the nature of the work and data quality requirements. In large routine analysis operations, where many similar samples are analyzed concurrently in a batch or run, one suitable spiked recovery sample will suffice to monitor the performance of the method for quite a number of samples.

4. DETERMINATION OF RECOVERY BASED ON NATIVE ANALYTE

4.1 Philosophy on Recovery of Native Analyte

The level of accuracy of trace element analysis depends greatly not only on the concentration of the analyte itself but also on the nature of the native analyte and the complex effects of interferences from sample matrix components. The term 'matrix' collectively encompasses all constituents, of the sample undergoing analysis, other than the analyte sought, such as ash, silica, fibre, protein, carbohydrate, fat and all other major (especially) and minor elements. The term 'native analyte' refers to the total analyte content, that is the sum of all forms and species of the element in all phases of the material. For example, 'total Kjeldahl nitrogen' refers to the total concentration of nitrogen in the material, that is all forms of nitrogen (in the protein, other organic, inorganic compounds) to the extent measurable by the Kjeldahl method, suitably adapted. It is incumbent on the analyst to incorporate whatever steps (oxidizing/reducing conditions and substances, temperature, etc.) deemed necessary and appropriate in the digestion and determinative phases to arrive at such analytical results. Confidence in the analytical method is substantially increased by knowing the method's performance for the analyte of interest with real-life samples with known levels of native analyte.

Thus, whereas acceptable recovery with spiked analytes is an essential criterion for satisfactory method performance, it by itself is deemed to be insufficient. One major thesis of our presentation is the necessity for estimating recovery of native, endogenous analyte, from natural, real matrix materials, in addition to demonstration of suitable performance with artificially - added analyte. Therefore one deals with **two levels or types** of recovery for more complete and essential characterization of method performance. Whereas, in the determination of organic analytes, recourse to spiking is, at times, the only option available to the analyst to estimate analyte recovery, for elemental determinations, an additional second option is the use of Reference Materials to arrive at another measure of recovery.

One type of interference mentioned in a preceding section is retention of the element by undissolved matrix components remaining after incomplete sample dissolution. Retention can be by simple physical adsorption or occlusion, chemisorption of the element onto active surfaces of the residue or actual chemical incorporation into the structure. In a previous study,[5] it was found that insoluble siliceous residues remaining after nitric / perchloric acid decomposition of three plant tissue Reference Materials contained varying quantities of macro and micro elements Na, K, Mg, Ca, Mn, Fe, Cu and Zn. For the different materials with total element concentrations ranging from 11 mg/kg to 45,000 mg/kg, residues contained element concentrations ranging from 0.05 to 88 times the respective concentrations in the samples, equivalent to 0.04 to 42 % of the total concentrations. The conclusion was that residue contributions to total element levels must be considered for reliable estimations of total element concentrations. It was recommended that the siliceous residue be dissolved with hydrofluoric acid either separately or in a one-step procedure, and that these residue-bound element concentrations be incorporated with the soluble fraction to correctly arrive at a determination of total concentrations. In another work with a Whole Egg Powder Reference Material[6], which has an added aluminosilicate anti-caking agent, the heavy insoluble siliceous residue was removed by filtration, dissolved in hydrofluoric acid, and analyzed separately; total concentrations were obtained by combining the results from both fractions. The many methodologies which advocate settling out of residue or its removal by filtration will therefore not measure total contents. This will be evident when recovery controls are conducted using natural analyte / natural matrix Reference Materials but will not be revealed using solely spike recovery studies.

Increasingly more and more effort is being directed to data quality, data intercomparability and harmonization of methods and results with the concomitant increase in reliance on use of biological and environmental Reference Materials as integral components of quality control. It would therefor seem appropriate, convenient and cost-effective to use results obtained with such materials for estimation of native recovery. Reference Materials are available for a wide range of nutritionally-, toxicologically- and environmentally- pertinent elements in varied matrices. They offer advantages to the calculation of recovery factors based on **native forms and levels** of analytes in natural matrix materials. Disadvantages and limitations are mainly due to unavailability of materials containing all of the analytes, levels and matrices which may be required in a specific case and the effect of uncertainties of certified values on the calculation of recovery factors as discussed below.

4.2 Procedures for Recovery Determinations Based on Native Analyte

Recovery of endogenous inorganic analytes from natural matrix commodities is easily feasible on account of the large and growing availability of a large number and wide variety of natural materials with defined elemental content. Suitable control materials can be chosen from Reference Materials, in-house quality control materials and proficiency testing materials. All afford some degree of control and different advantages and disadvantages.

A Reference Material is defined as a material with one or more properties sufficiently well established or characterized to be used for the calibration of an apparatus, the assessment of a measurement method, or for assignment of numerical property values to materials. A Certified Reference Material is defined as a Reference Material one or more of whose property values are certified by a technically-valid procedure, accompanied by or traceable to a certificate or other documentation which is issued by a certifying body. The term certification refers to the assignment of reliable numerical concentration values to Reference Materials, generally backed by legal mandate, under the auspices of a national or international agency with demonstrated technical competence in the standards field. Characterization has a more general connotation referring to concentration assignments resulting from work outside such agencies. Either of these types of Reference Materials, referred to in this paper as Reference Materials, with well established, good, solid, reliable concentration values are deemed to be acceptable as bone fide Reference Materials for recovery determination.

In-house quality control and proficiency testing materials belong to a similar class of natural products. As the names suggest, these materials have been developed within the laboratory for day-to-day quality control activities or by an outside laboratory for submission to client laboratories for proficiency testing. Although, strictly, such products have not necessarily been accurately characterized, they often have respectably precise values and have a role to play in recovery testing.

Considerations for utilization of natural matrix control materials for recovery measurement are in many respects similar to those discussed above for procedures for recovery determination using spiking. Again, consideration of determination of recovery for the different matrix types encountered in analysis or only for selected representative matrices is a function of the importance of the analyses. A preliminary semiquantitative analysis of selected samples would be advantageous to get a ballpark value of analyte level and to establish at least an estimate of matrix composition to facilitate selection of the most representative and suitable Reference or control materials from those available.

For correct and effective use of a Reference Material, the material selected must resemble, in all respects, as closely as possible, the actual materials being analyzed. It must be very similar with respect to matrix and analyte (concentration level and form eg. native form, speciation) to the commodity undergoing analysis. Selection can be based on information in Reference Material catalogues and in other publications[7-11] issued as guides for the analyst, containing useful, detailed compilations of materials, analyte concentrations, and suppliers, facilitating selection of appropriate materials. For multielement analyses, that is the determination of more than one element on the same laboratory subsample, identical selection steps are pursued for the second, third and other elements to choose appropriate materials for each of these respective analytes. It is advisable to maximize the number of elements to monitor by a given Reference Material and minimize the number of materials required, by reducing the strictness of matrix and analyte matching criteria. Selection criteria rigidity is at the analyst's discretion and is governed by the level of quality control desired, the availability of Reference Materials and the rate of Reference Material incorporation deemed acceptable for recovery measurements.

Analyte levels should be selected to represent the expected level of analyte to be measured or to approximate enforcement / quantitation / determination / reporting limits. As with spiked samples due consideration should be given to ensuring that the instrumental analytical response for the Reference Material falls on the calibration scale to permit adherence to identical conditions of dilution, calibration etc. as for actual sample. A sufficient number of repeat measurements at each concentration level should be carried out in order to get a good estimate of measurement uncertainty (precision). Unlike the case of spiking, assessment of amount of analyte already present in material (baseline level) is not required prior to conduction of recovery tests (however, see below) as the baseline value is simply the certified or recommended concentration value. The rate of incorporation of Reference Materials could range from less than 1/100 samples to more than 1/10 samples, depending on the nature of the work and data quality requirements. In large routine analysis operations, where many similar samples are analyzed concurrently, one suitable Reference Material will suffice to monitor the performance of the method for quite a number of samples.

For proper utilization, instructions for material usage and handling set out in the certificate / report of analysis are followed and the Reference Material is incorporated into the scheme of analysis at the earliest stage possible, ie. prior to the beginning of sample decomposition. It is taken through the entire analytical procedure at the same time and under the identical conditions as the actual analytical samples in order to correctly monitor all the sample manipulation and measurement steps. For multielement determinations, should different sample preparation and measurement procedures (ie. different analytical methods) be indicated for the different elements, separate aliquots of the original Reference Material are taken through the entire relevant analytical scheme for proper quality control.

An advantage of using Reference, in-house quality control and proficiency test materials is that the level of the native analyte therein, as well as the matrix constituents are at, or close to, respective natural levels and the recovery test with such a material measures recovery of the real, native analyte at a real-life level in a real-life, natural matrix material. Additionally, Reference Materials have well known, certified analyte levels for calculating recovery factors. Natural in-house and proficiency control materials may possess other additional advantages over Reference Materials. Their composition (especially that of in-house control products selected or developed specifically to be quite similar to samples undergoing analysis) may be more similar to the actual samples and thus their behaviour may more closely mimic that of actual samples giving a better indication of recovery. They also may be more homogeneous, particularly when occurring as liquids, slurries or semisolid pastes.

One disadvantage to relying solely on Reference Materials is the limited availability of Reference Materials in the world repertoire reflecting analytes of interest, levels required, and matrices desired. A second disadvantage, if it may be termed so and to be discussed in depth below and in the section on calculation of recovery, is the unavoidably large uncertainty generally associated with every certified or recommended concentration value and its contribution to the uncertainty of the final recovery factor.

The large uncertainties in certified and recommended concentration values contribute substantially to uncertainties in calculated recovery factors. Uncertainties reported for certified or recommended elemental concentrations result from material inhomogeneity and certification measurement imprecisions and biases, and are generally several fold greater than the uncertainties of analytical determinations usually encountered in a single laboratory using a single method. This stems from, particularly, the incorporation of data in Reference Material characterization from different methods/analysts with different biases and systematic errors, by necessity, inflating the overall uncertainty (but nevertheless resulting in <u>accurate</u> data). Such multi-method and inter-laboratory biases do not come into play when single-method, single-laboratory determinations are conducted. To reduce the magnitude of errors, it would be appropriate and most suitable to use concentration values obtained by the identical method used in the laboratory for recovery tests. Such information is invariably not available in certificates of analysis, and in our view, not to be reported in such documents (as done in the case of some clinical work). Alternatively, one can arrive at a good estimate of this value, containing only precision errors, to use as a baseline, although at some laboratory labour, by repetitive complete analyses of the Reference Material using the one method used in the laboratory. This will result in a baseline estimate with a good, low precision, contributing much less error to the final calculated recovery factor. It is advisable, however, that such a material be used only for recovery tests determining recovery above the 'baseline' level of a spiked analyte. As a quantity of expensive Reference Material will be consumed by this type of recovery application it would be preferable to apply this principle to recovery testing using in-house and proficiency control materials as well.

4.3 Blank Materials

Determination of the blank in analytical chemistry is an integral and vital component of every analytical method and determination including recovery. The blank value for the analyte of interest, in principle, corrects for all occurrences in the execution of the method that are to be subtracted from the gross analyte value to arrive at a correct estimate of the analyte concentration. A blank value is typically determined by taking the chemical reagents (omitting the sample) through the entire method, applied in a manner as close as possible to that with sample present. In spite of attempts at duplicating the procedure in blank determinations, performance of the method with reagents alone may not always be exactly identical to performance with sample present. For example, presence of natural material throughout the decomposition stage may result in more attack on the decomposition vessels and consequently more chance for extraction of analyte therefrom as a contaminant than would be the case with reagents alone. A reagent blank determined in this manner would underestimate the real blank contribution and hence would not provide for accurate blank control or correction. Similarly, presence of material throughout the decomposition stage may be responsible for more or less retention of analyte by the decomposition vessels than would be the case with reagents alone. A reagent blank would, respectively, underestimate or overestimate the real blank and hence again would not provide for proper control or correction.

Availability and incorporation of a 'real' blank sample would overcome, to a large degree, such differences in behaviour between reagent blanks and samples and provide for superior measures of sample ashing or digestion blank values and correct blank control. A 'real matrix blank' material can be defined as any native analyte/natural matrix material with elemental concentrations substantially below the limit of detection of the method used and of composition appropriate to the task. In the absence of blank effects, the method would produce a value of 0. Typically a non-zero value will result which would be construed as the blank. Obviously, the concentration of the analyte in the blank material must be known with certainty (so that the analyst can in fact confidently take it to be zero for his application) and thus a Reference Material is the natural choice. Perusal of elemental concentration tables, especially those in order of concentration,[7-9,11] can indicate which materials have low element levels and can conceivably be considered for use as real blank control materials in some analytical procedures. A limited number of such 'blank Reference Material' products is available, including NIST RM 8416 Microcrystalline Cellulose and NIST RM 8432 Corn Starch which contain very low levels for a good number of elements. In addition to the establishment of the zero value in absence of native and added analyte, 'blank Reference Materials' are also valuable for determining recovery when no native analyte is present. Determination of recovery of spiked analyte from a low level (essentially zero level for certain procedures) matrix sample allows for a high ratio of added to native analyte and thus a good estimate of recovery with incursion of smaller baseline errors due to the very low starting analyte content.

5. CALCULATION OF RECOVERY

5.1 Calculation of Recovery Values

The recovery factor is a quantitative representation of the proportion of added or endogenous analyte recovered and measured by the specified overall method and defines one performance aspect, perhaps the most important one, of the method. The factor is expressed as the ratio of analyte recovered to analyte added or known to be present and is given as a numerical value with an associated uncertainty. Ideally it has a numerical value of 1.000 but in reality will deviate positively or negatively from unity and is to be reported, as determined, as a number greater or less than 1. Measurements and calculations are separately conducted for added or native analyte and the two are not mixed. For example, when recovery of added analyte is determined, correction is made for the background content of native analyte so that the recovery factor reflects performance solely with added analyte. The practice of calculating 'blended recoveries' where the issue of spike and native recovery is confounded is not advocated.

Generally, two independent determinations of concentration constitute the basis for the calculation of the recovery factor. For calculations of added analyte (spiking), the first measurement is the analysis of the sample without added analyte to establish the baseline value. The second analysis is of the sample with the spike. The factor is calculated from:

$$F = [C_{b+s} - C_b] / C_s \qquad\qquad (1)$$

Where:

F is the recovery factor,

C_{b+s} is the concentration of the analyte determined in the spiked sample,

C_b is the concentration of the analyte in the sample (baseline content),

C_s is the theoretically expected concentration of the added analyte, due to the spike, in the sample.

This determination and calculation indicates the performance of the method with respect to the added analyte only and yields a value for the recovery of the added element independently of the performance of the method with native analyte.

For calculations of native analyte using Reference, in-house, or proficiency control materials, there is generally no requirement to conduct the first measurement of baseline content (at least not at the time of recovery determination). This first measurement has already been done either by the producer in the case of Reference Materials or at some time previously by the analyst for the case of in-house or proficiency control materials. For Reference Materials this value is obtained from information in the certificate or report of analysis. An example of a Table from a Report of Investigation[12] listing best estimate concentrations of constituent elements, associated uncertainties and methods used in certification, is given in Table 1. Analysis of the material by the analyst for recovery constitutes the second analysis. The recovery factor is calculated from

$$F = C_{analyst} / C_{reference} \qquad (4)$$

Where:

$C_{analyst}$ is the concentration of the analyte in the control material, determined by the analyst, $C_{reference}$ is the certified, recommended or otherwise appropriately determined concentration of the analyte in the sample.

The value $C_{reference}$ refers to the certified/recommended value reported in the certificate or report of analysis in the case of Reference Materials, or an otherwise appropriately determined value for the Reference, in-house or proficiency control material accepted by the analyst as a suitable value for recovery calculation.

As with all analytical readings and results, the question of outliers arises when dealing with data leading to recovery factors and the factors themselves. It is our belief and contention that only obviously erroneous data, especially those for which an explanation is apparent, be omitted. Otherwise, the temptation to delete aberrant or outlying data should not be pursued lest such rejection put the performance of the method in a more favourable light than justified. Generally, essentially all recovery control results should be retained. That being accepted means that minimal reliance need be placed on implicit or statistical protocols for data evaluation and calculation of recovery factors.

5.2 Estimating Uncertainty of Recovery Values

In line with good analytical and measurement practice, where an estimate should always be reported with every datum, the recovery factor should be given an associated uncertainty. Uncertainty is defined as any appropriate combination of precision and bias/systematic errors, in all the required determinations or giving an estimate or indication of the overall possible error of the recovery. Uncertainty calculations can be based on laboratory - determined, laboratory-known precision and bias/systematic errors. Appropriate statistical calculations in calculating uncertainties associated with the recovery factors. Since dependent on concentration (as aptly predicted by Margosis et al.,[ ] relation (Horwitz curve) Recovery precision and concentration) approximation, independent of method as well as other parameters, it at the various concentration levels encountered for more reliable estimates of recovery in turn.

In the case of spike recoveries from samples, repetitive background (background spiked concentration, $C_{}$) will provide definition) estimates for the two required determinations which will estimates for the recovery factor. Precision and systematic errors expected concentration of the added analyte, is expected to be and purposes, except the most rigorous endeavour, may be ignored. Error in C_s comes basically from (a) reliability of spike solution preparation and (b) reliability of spike addition technique when introducing it into the sample. Error in C_b depends on material homogeneity and the analytical method while error in C_{b+s} depends on material homogeneity, analytical method and error in C_s. Propagation of error formulas may be utilized to calculate the transmission of accumulated error to the final recovery factor. Much more mathematically complex uncertainty computations will have to be resorted to should the concentrations be considered as dependent variables, instead of independent variables. Identical computations are followed for the similar cases of recovery

Table 1 *Example of a Table from a Report of Investigation, Listing Best Estimate Concentrations of Constituent Elements, Associated Uncertainties and Methods Used in Certification, for Reference Material Wheat Gluten WG 184 (NIST RM 8418)*

Major Constituents

Element	*Content and uncertainty* Weight %(a)	*Methods(b)*
Nitrogen	14.68 ± 0.26	101 I02 J01 J02
Sulfur	0.845 ± 0.085	B02 B03 F04 J02 M02
Chlorine	0.362 ± 0.022	D01 F02 K01 K02
Phosphorus	0.219 ± 0.015	B02 B03 F01 F02 M01
Sodium	0.142 ± 0.011	A01 B01 B02 D01

Minor and Trace Constituents

Element	*Content and uncertainty* mg/kg(a)	*Methods(b)*
Magnesium	510 ± 47	A01 B02 B03 D01
Potassium	472 ± 61	A01 B02 B03 D01 E01
Calcium	369 ± 35	A01 B02 B03 D01 E02
Iron	54.3 ± 6.8	A01 B02 B03 D01 D03 E01 E02
Zinc	53.8 ± 3.7	A01 B02 B03 D03 E01
Manganese	14.3 ± 0.8	A01 B02 B04 D01 E01 E02
Aluminium	10.8 ± 3.0	A05 B02 B03 D01
Copper	5.94 ± 0.72	A01 A05 B02 C03 C06 E01 H01
Selenium	2.58 ± 0.19	B02 C01 C04 D01 D03 G01
Strontium	1.71 ± 0.26	B02 B03 C03 E01
Barium	1.53 ± 0.26	B02 B03 C03
Molybdenum	0.76 ± 0.09	B02 C03 C06 D01 D03 F01 H06
Nickel	0.13 ± 0.04	A16 H01
Lead	0.10 ± 0.05	A05 A16 C03 H01
Cadmium	0.064 ± 0.022	A04 A05 A16 C03 D03 H01
Iodine	0.060 ± 0.013	D03 D05 D06 F02 H03
Chromium	0.053 ± 0.013	A12 C05 D03
Cobalt	0.010 ± 0.006	A16 D01 H01
Mercury	0.0019 ± 0.0006	A10 D03

(a) Best estimate values are equally-weighted means of results from typically several, different analytical methods applied by analysts in different laboratories. Uncertainties are estimates expressed either as a 95% confidence interval or occasionally as an interval based on the range of accepted results and are based on among-method and laboratory, among-unit and within-unit estimates of variances; refer to the report 12 for details.

(b) Analytical method codes and descriptions are in the original report 12.

of added analyte from Reference, in-house or proficiency materials, where background and spiked concentrations are measured by repetitive determinations on the control and spiked control materials. If only determination of C_{b+s} is necessary and C_b is taken from the certificate of analysis, then precision of C_{b+s} and uncertainty in the actual, certified level from the certificate of analysis are the two uncertainty components.

As an example for the case of spike recovery, if we assume the following values and errors (say standard deviations) for each of the variables used in calculating F from eqn: 1, C_b = 2 ± 0.2 mg/kg (RSD = 10 %), C_{b+s} = 12 ± 0.4 mg/kg (3.2%), C_s = 10 ± 0 mg/kg (0 %), the recovery factor, F = 1.000 ± 0.045 (4.5 %). In this case it is assumed that the baseline level is determined to ± 10 %, and the higher level spiked sample to ± 3.3 %; the error in the theoretically expected concentration of the added spike is taken to be negligible and assigned a value of 0 %. Further, for demonstration purposes F is taken to be unity.

In the case of native analyte recoveries using Reference Materials, repetitive determinations on the control material, generally the one required determination, will provide a precision estimate of $C_{analyst}$. The uncertainty in the actual, certified level, $C_{reference}$ is taken from the certificate of analysis and these two uncertainties lead to uncertainty estimates in the recovery factor. Error in $C_{analyst}$ depends on material homogeneity, analytical method/analyst while error in $C_{reference}$ depends on material homogeneity, analytical method/analyst, interlaboratory systematic errors and other considerations, introduced into the uncertainty, by the Reference Material-producing agency as discussed in sect. 4.2. Application of appropriate propagation of error formulas again may be used to calculate the transmission of accumulated error to the final recovery factor.

Again, should the concentrations be considered as dependent variables, resort will have to made to much more mathematically complex uncertainty computations. For the similar case of native analyte recoveries using Reference, in-house or proficiency materials, based on $C_{reference}$ established by the analyst by previous repetitive determinations, on the control material using the laboratory method, the one required determination (of $C_{analyst}$), will provide a precision estimate of $C_{analyst}$. The uncertainty in the reference level established in the laboratory is taken from the determined precision rather than the certificate of analysis.

As an example for the case of recovery of native analyte, say sulfur, from Reference Material Wheat Gluten, NIST RM 8418, if we assume the following values and errors (say standard deviations for analyses in the laboratory, uncertainties as reported in certificates of analysis) for each of the variables used in calculating F from eqn: 2, $C_{analyst}$ = 0.854 ± 0.024 % (3.0 %), $C_{reference}$ = 0.854 ± 0.085 % (10.0 %), the recovery factor, F = 1.000 ± 0.105 (10.5 %). We note that the uncertainty of analysis conducted by the analyst (3 %) is taken to be lower than the uncertainty in the reference value (10 %), taken from Table 1, as is usually the case.

As an example for recovery of native cadmium, present at a much lower concentration of 0.064 mg/kg, from the same Reference Material, if we assume the following values and errors for each of the variables used in calculating F from eqn: 2, $C_{analyst} = 0.064 \pm 0.0064$ mg/kg (10.0 %), $C_{reference} = 0.064 \pm 0.022$ mg/kg (34.0 %), the recovery factor, F = 1.000 $\pm$ 0.358 (35.8 %). In this case it is assumed that the level determined by the analyst is determined to 10 %; the error in the reference level at 34 % is taken from Table 1. In both the sulfur and cadmium instances, it may be noted that the uncertainty in the reference value, being much larger than that of the value determined by the analyst, makes an overwhelming contribution the uncertainty of the recovery factor.

When considering the formal calculation of native analyte recovery factor uncertainties, using uncertainties reported for Reference, in house or proficiency control materials, the use of such materials may be, in a fashion, considered a disadvantage. Uncertainties reported for certified or recommended elemental concentrations result from material inhomogeneity and certification measurement imprecisions and biases, and are generally several-fold greater than the uncertainties of analytical determinations usually encountered in a single laboratory using a single method. This stems from, particularly, the incorporation of data, from different methods/analysts with different biases and systematic errors, in control material characterization. By necessity, this inflates the overall uncertainty (but nevertheless results in accurate data). Incorporation of such large uncertainties will have a substantial impact on the calculated uncertainty of the recovery factor, making the magnitude of the uncertainty fairly large. Reference Material uncertainties are obtained from information in the certificate of analysis such as, for example, Table 1.

In Reference Material characterization, attempts are purposely made to get analyses by wide-ranging techniques and procedures, including different sample preparation steps, including no decomposition as in instrumental neutron activation analysis and particle induced X-ray emission spectrometry, as well as different detection and measurement techniques. The philosophy behind this is that inclusion of data from diverse, independent methods, applied by different analysts, entailing, in principle and in practice, different biases, is an important consideration in arriving at the best estimate of the 'true value'. The large variety of sample treatment, ashing and digestion approaches as well as detection/measurement schemes and the many variants thereof and other safety factor considerations of the producer inflate the overall uncertainties of the certified values.

For data in Table 1, best estimate values, weight percent or mg/kg (ppm), are based on the dry material, dried according to instructions in the report and are equally-weighted means of results from generally at least two, but typically several, different analytical methods applied by analysts in different laboratories. Uncertainties are estimates expressed either as a 95% confidence interval or occasionally (Co, S, Se) as an interval based on the range of accepted results for a single future determination based on a subsample weight of at least 0.5 g. These uncertainties, based on among-method and laboratory, among-unit and within-unit estimates of variances, include measures of analytical method and laboratory imprecisions and biases and material inhomogeneity.[12]

5.3 Measurement/Calculation of Recovery in Chromatographic Analysis

One of the fallacies common among analytical chemists is that the lower the blank, the more accurate the results. The major exception to the above statement is instrumental treatment of response information encountered with the use of chromatographic equipment as well as emission spectrometers with automated peak identification software. This refers to those systems which do not make a measurement in a given spectral or chromatographic window unless there is an actual peak present. All systems have some degree of baseline noise. These systems are designed to identify and measure a response peak only if it is some factor above the baseline noise. The problem with such measurement systems is that if a peak occurs between the baseline and the lower boundary set for peak identification, then, instrumentally, the concentration will be assumed to be zero. Also, chromatographic software frequently has serious difficulty with the correct measurement of baselines. This raises several problems, depending on the solution generating the peak.

In the simplest case, if a substantial proportion of samples being measured have concentrations which are near the cut-off point of the identification software, the distribution of concentrations from the samples will be skewed; i.e., there will be a large number of concentrations above the cut-off point as well as a large number of zero readings. The latter will result in an analytical mean which is biased toward lower concentrations.

In a more complex case, if the solutions being measured are reagent blanks and the concentrations are below the cut-off point but above zero, then the blank which is subtracted from sample concentrations will be defined as zero. This is an invalid blank and results in sample concentrations biased toward higher concentrations. It also results in a situation where it is impossible to monitor the true standard deviation of the blank in order to assess both the detection limit and the presence or absence of adventitious contamination.

Neither of the above biases can be identified using recovery studies irrespective of spiking level. here are only two ways to overcome these biases. He first is to choose software which integrates a window defined by the standards regardless of whether or not a peak is identified; i.e., eliminate the peak identification algorithm for all samples; this would require major revision to the software. The second approach is to make sure that the peaks are identified for each and every solution, including the blanks. This can be done by actually spiking all the solutions, including the blanks, standards, samples, sample spikes, reference materials, etc., with a small amount of the analyte measured. The amount added should be just sufficient to give identifiable peaks. This goes contrary to what is normally assumed to be a rule in analytical chemistry: the lower the blank, the more accurate the analysis. In the presence of software of this type, a positive blank is not only desirable, it is necessary for accuracy.

6. DIFFERENTIATION BETWEEN RECOVERY AND BIAS/SYSTEMATIC ERROR

One of the major misconceptions of less experienced analysts is that good recovery means good accuracy. Recovery studies are the most frequently used method of sample result or method validation. They do not reflect analytical accuracy, however, because they only evaluate recovery of the analyte added to the sample and tell us nothing about the amount of analyte present in the sample. That is, they give no indication about the accuracy of the unspiked sample signal. Thus, recovery can be 100%, yet analytical results can be biased and in error by orders of magnitude.

Situations causing such errors are (a) contamination of samples but not blanks, (b) contamination of blanks but not samples, (c) presence of uncorrected background in atomic absorption spectrometry contributing to a portion or to all of the analyte signal, or (d) an invalid baseline in chromatography or stripping voltammetry. For example, when foods are dry-ashed for lead analysis in quartz or Pyrex vessels, lead present in the surface of the interior of the vessel can be leached into the sample ash by the aggressive nature of some of the ash components. Because the blank has no sample ash present, the measured reagent blank will be artificially low, and sample concentrations will be biased toward higher concentrations.

The ways to evaluate the presence of any of the above errors are to include appropriate Reference Materials with baseline-levels of analyte (refer to blank mention above), or to analyze the samples using a completely independent method of analysis. The former is infrequently used because of availability and cost of good control materials and even when Reference Materials are used, analysts prefer to choose those with higher concentrations because the quality control results "look better". The use of an independent method of analysis is usually impractical due to productivity demands on the analyst.

A sample weight test which overcomes the above limitations with little additional time or cost to the analyst has been described recently.14 The test involves analyzing two different weights of the same sample. One weight should be at least twice that of the other. The number of replicates determined at each sample weight depends on the critical nature of the sample, the homogeneity of the sample and the precision of the method.

If the same analyte concentration is obtained for the two different sample weights, then the result can be considered accurate. If different concentrations are obtained for the two sample weights, then the results should not be reported and a cause for the discrepancy should be sought. The sample weight test is particularly sensitive at low concentrations, and can reveal most of the method bias problems mentioned above. Thus, it is complementary to recovery studies as an evaluation of accuracy. The test should be applied to all samples of a critical nature (sample result validation) and to all test samples when a method is being validated.

Also relevant is the question of differences in the nature and extent of interactions of analyte with sample matrix referred to in sect. 3.2. Complex kinetically- and thermodynamically-driven interactions can occur including intra-particle diffusion, physical and chemical binding, precipitation and other phenomena making the analyte unavailable to subsequent detection. The resolution of this issue would be easier if one had more understanding of such interactions.

7. APPLICATION OF RECOVERY FACTORS

Recovery study results should be used with caution. As a general rule, recovery is only used to assess the performance of the method with a particular sample. If a numerical value differing from unity and either greater or less is obtained for the recovery factor, a discrepancy is deemed to exist between the measured and correct concentration value indicating the analytical procedure not to be operating well. Should it be ascertained that an unacceptable error exists, a correction should not generally be applied. Instead, diagnostic steps should be taken to identify sources of unacceptable error or imprecision and remedial action should be taken to eliminate or at least minimize such errors in the method. It is our contention that recovery factors generally should not be used to adjust the results to correct for recovery.

Our philosophy is thus: estimating, measuring, determining recoveries of elemental analytes, not using such factors for correction but rather using them as a flag of method performance and an indication for the need to search out sources of error and eliminate or at least reduce magnitudes of random and especially systematic errors. No compensation is made for the loss or gain of analyte in calculation of final concentration and the concentration is reported uncorrected or unadjusted for recovery, with an indication of the recovery factor included in the report. The aim is to have quantitative or sufficiently high, good recovery to be able to ignore the factor (ie equate to 1). The recovery factor, measured throughout the various stages of method fine tuning, development and application, serves to track method performance during development with the goal of arriving at a method with quantitative or sufficiently acceptable performance.

Having stated that, there are circumstances when it is valid to use recovery studies to adjust sample concentration results for losses or enhancement by application of the recovery factor, and we feel that such adjustment is <u>sometimes</u> justified. For example, if a method is well defined and used by an experienced analyst, it is known whether or not the method has a bias. If the method bias is directly proportional to recovery, and this is known without a doubt, then it is valid, when high accuracy is needed, to use the recovery obtained in careful spiking studies to adjust the analytical results and thereby correct for recovery. When such corrections are made, however, it should be realized that the operation is defined as internal standardization rather than a recovery study, and the quotation of the recovery study results as part of quality control is invalid.

8. CONCLUDING REMARKS

A consideration has been presented of several of the many parameters which impact on the determination and utilization of recovery factors. The major points put forward are: (1) the necessity for estimating recovery of native, endogenous analyte, from natural, real matrix materials, in addition to demonstration of suitable performance with artificially - added analyte; (2) incorporation of a 'real' blank sample to overcome differences in behaviour between reagent blanks and samples and provide for superior measures of sample decomposition blank values and correct blank control as well as indicating possible method biases; (3) estimation and inclusion of an uncertainty with each recovery factor; (4) surmounting biases in measurements introduced by instrumental treatment of response information in chromatographic and emission spectrometric determinations incorporating automated peak identification and measurement software; (5) our philosophy for utilization of recovery factors as a measure of method performance and an indication for the need to search out and eliminate or reduce errors rather than use of such factors for correction of analytical results. There are additional ramifications of these parameters as well as other facets which play a role and must be considered, debated and evaluated in an effort to arrive at a harmonization of philosophy, concepts, approaches and best practices for the measurement and use of recovery factors. Hopefully the ideas presented here will stimulate discussion toward the quest for a standardized practice for the measurement and use of recovery factors.

9. REFERENCES

1. V. Iyengar, *Anal. Chem.*, 1982, **54**, 554A.
2. B. Kratochvil and J.K. Taylor, *Anal. Chem.*, 1981, **53**, 924A.
3. J. Versieck, F. Barbier, R. Cornelis and J. Hoste, *Talanta*, 1982, **29**, 973.
4. J. Versieck, *Trace Elements Medicine*, 1984, **1**, 2.
5. M. Ihnat, *Comm. Soil Sci. Plant Anal.*, 1982, **13**, 969.
6. M. Ihnat, *Fresenius' J. Anal. Chem.*, 1993, **345**, 221.
7. Y. Muramatsu and R.M. Parr, 'Survey of Currently Available Reference Materials for use in Connection with the Determination of Trace Elements in Biological and Environmental Materials', IAEA/RL/128, International Atomic Energy Agency, Vienna, 1985.
8. M. Ihnat, *in* 'Quantitative Trace Analysis of Biological Materials', H.A. McKenzie and L.E. Smythe, eds., Elsevier, Amsterdam, 1988, Appendix 1, p. 739.
9. E. Cortes Toro, R.M. Parr and S.A. Clements, 'Biological and Environmental Reference Materials for Trace Elements, Nuclides and Organic Microcontaminants', IAEA/RL/128 (Rev. 1), International Atomic Energy Agency, Vienna, 1990.
10. A.Y. Cantillo, 'Standard and Reference Materials for Marine Science', 3rd ed., NOAA Technical Memorandum NOS ORCA 68, National Oceanic and Atmospheric Administration, Rockville, MD, 1992.
11. M. Ihnat, in 'Soil Sampling and Methods of Analysis,' M.R. Carter, ed., Lewis Publishers, Boca Raton, 1993, p. 247.
12. M. Ihnat, 'Report of Investigation, Reference Material 8418 Wheat Gluten', National Institute of Standards and Technology, Gaithersburg, MD, 1993; also: 'Report of Investigation, Reference Material WG 184 (NIST RM 8418) Wheat Gluten', Centre for Land and Biological Resources Research, Agriculture Canada, Ottawa, ON, 1993, 15pp.
13. M. Margosis, W. Horwitz and R. Albert, J. Assoc. Offic. Anal. Chem., 1988, **71**, 619.
14. R.W. Dabeka and S. Hayward, *in* 'Quality Assurance for Analytical Laboratories', M. Parkany, ed., Royal Society of Chemistry, London, 1993, p. 67.

Contribution no. 96-04 from Centre for Land and Biological Resources Research

Use of Radioisotopic Tracers for Determination of Recovery Factors (Chemical Yields) in Trace Element Determinations by Radiochemical Neutron Activation Analysis (RNAA)

A. Fajgelj[1] and A. R. Byrne[2]

[1]INTERNATIONAL ATOMIC ENERGY AGENCY LABORATORIES, A-2444 SEIBERSDORF, AUSTRIA

[2]J. STEFAN INSTITUTE, 61111 LJUBLJANA, SLOVENIA

1. INTRODUCTION

The versatility and applicability of neutron activation analysis (NAA) for determination of trace element contents in different types of samples (environmental, biological, medical, forensic, etc.) is very well known. It has also been shown that NAA can even be applied in some speciation studies.[1,2] The advantages of NAA are, in brief: sensitivity, the virtual absence of an analytical blank, relative freedom from matrix and interference effects, its nondestructive nature in many cases (instrumental neutron activation analysis INAA), its property in its radiochemical mode (radiochemical neutron activation analysis - RNAA of allowing trace element radiochemistry to be performed at higher and controlled conditions by nonradioactive carrier addition, different possibilities of chemical yield (recovery) determination, its high specificity, its totally independent principle as a nuclear method, and its isotopic basis. Taking into account only the relative approach for calculation of the mass of element(s) determined, where a standard (normally a solution or a dried aliquot thereof) of this element is neutron irradiated simultaneously with the sample and measured under the same measuring conditions as the sample, the method is also theoretically simple and well understood. This means that sources of uncertainty can be easily defined and uncertainty components evaluated or at least estimated.[3, 4] The use of high purity chemical irradiation standards (prepared from pure metals or stoichiometrically well defined chemical compounds)[5], a good description of the method (easy transfer to mathematical formulae) and evaluation of uncertainty are the basic requirements for establishing the traceability of quantitative chemical measurement to SI units.[6]

In this paper, we intend to stress the important contribution of the chemical yield to the quality of RNAA, and the advantages of radiotracers for its determination.

2. PROCEDURES IN RNAA

An analytical procedure in RNAA should normally involve eight steps: i) sample and standard(s) preparation, ii) neutron irradiation, iii) addition of carriers, or radioisotopic tracer, or both, iv) sample + carrier dissolution, v) radiochemical separation, vi) radioactivity measurement, vii) yield (recovery) determination, viii) calculation of result(s). The main source of uncertainty in RNAA is often the chemical yield. The chemical yield will never be precisely reproducible, and should be measured in each analysis.[7]

When a radiochemical procedure is developed, the chemical yield and its variability is an important characteristic which should be investigated for different matrices. Yields are usually established at the development stage by the tracer method. Even when the yield is apparently quantitative, it should still be measured in analysis of each sample aliquot to ensure quality control of the result; human error, losses from unforeseen causes, contamination, etc. can never be ruled out.

One of the following three methods (all based on the fact that different isotopes of the same element behave chemically in the same way when in the same oxidation state and chemical form) is usually used for this purpose:

a) Carrier added to the sample before digestion (macro amount of the element determined) can be used for recovery determination by conventional analytical techniques (gravimetry, colorimetry, etc.).

b) After measuring the radioactivity induced by the original activation of the sample, the separated fraction (carriers together with the trace elements from the sample) is reactivated by irradiation with neutrons, and the amount of the carrier(s) recovered is determined by INAA (the contribution to induced radioactivity from the element(s) originally presented in the sample is negligible - trace amounts). This is the re-activation method.

c) The most direct method is the use of radioactive tracers, added with or without carrier(s) to the sample before digestion. During the radioactivity measurement (see vi) the activity of the tracer is also measured for recovery determination. This combines steps vi and vii into one step. Thus, the yield measurement procedure is not only simplified by one stage and efficient, but also increased in accuracy by compensating for factors affecting the count such as measuring geometry, pile-up, or dead time losses.[4] In comparison with the two other techniques (a and b) this also minimize the sources of uncertainty accompanying additional preparation and measurement of sample and standards by classical methods (gravimetry, colorimetry, etc. - see examples in ref. 3), or by additional neutron irradiation for recovery determination.

It is important to point out that a basic principle governing all work with tracers and isotopic methods in general, is the achievement of isotopic exchange as soon as possible in the procedure. In some cases, particularly with volatile elements such as Hg and I, this may not be simple, since part of the element may be in the gas phase after irradiation. Therefore freezing of sample in liquid nitrogen and addition of an oxidizing solution of the tracer or carrier directly to the sample in its irradiation vial is recommended.

3. REQUIREMENTS FOR APPROPRIATE USE OF RADIOACTIVE TRACERS AND SOME PRACTICAL EXAMPLES

The ideal or desirable characteristics of radioactive tracers to be practically applicable in chemical yield determinations include a lower gamma energy than the indicator peak (to avoid deteriorating the signal/background ratio), a longer half-life, and nonproduction (or negligible production) in the irradiated sample.[8, 9] With the development of low-energy photon detectors and well-type Ge detectors, which are more sensitive to lower-energy gamma-rays (i.e., 20-60 keV) than conventional Ge detectors, the range of radioactive tracers useful for chemical yield determinations has expanded. Some examples of radioactive tracers applicable for recovery determinations are listed in Table 1.

Table 1 *Some radioactive tracers (RAT) useful for recovery determination for trace element determination via indicator radionuclides (IRN) using RNAA.*

Element	IRN (g energy, keV)	RAT (g energy, keV)	Reference
Arsenic	^{76}As (559, 657)	^{73}As (58) ^{74}As (596, 635) ^{77}As (239)	10, 11
Cadmium	^{115m}Cd ® ^{115m}In (336) ^{109}Cd (88)		12
Cobalt	^{60}Co (1173, 1332)	^{57}Co (122 or 136) ^{56}Co (847, 1238 or 2598 keV)7	13
Copper	^{64}Cu (511)	^{67}Cu (184)	13
Iodine	^{128}I (443, 527)	^{123}I (159) ^{125}I (27) ^{131}I (364)	7
Lead	^{203}Pb (71, 73, 82.5, 279)	^{210}Pb (46.5)	12
Nickel	^{58}Co (811)	^{57}Co (122)	14
Palladium	^{109}Pd (88)	^{103}Pd (40, 397)	8
Selenium	^{75}Se (121, 136, 264, 279 400)	^{81m}Se (103)	15
Silver	^{110m}Ag (658, 885)	^{111}Ag (342) ^{108m} Ag (76, 434, 614, 723)	16 8
Thallium	^{202}Tl (70, 71, 80, 439.6)	^{201}Tl (135, 167) ^{204}Tl (69, 71, 80)	12

Table 1 - cont'd

Element	IRN (g energy, keV)	RAT (g energy, keV)	Reference
Thorium	^{233}Pa (233)	^{231}Pa (27 or 287)	17
Tin	^{117m}Sn (162)	^{113m}Sn (393)® ^{113m}In	8
Tungsten	^{187}W (480, 686)	^{181}W (56, 65)	16
		^{185}W (125)	8
Uranium	^{239}U (75)	^{235}U (185)	19
	^{239}Np (106, 228, 278)	^{238}Np (984, 1026, 1028)	18, 19
Zirconium	^{95}Zr (757)	^{88}Zr (393)	7
	^{97}Zr (508 or 1148 keV)		

As regards inappropriate use of tracers in RNAA, the practice (occasionally described) of processing the irradiation standard in the same manner as the samples must be condemned. Instead of the induced radionuclide in the standard performing its function of a direct calibrant, it is effectively used as a chemical yield monitor, thus subjecting it to similar but unknown yield variations as the samples. Even in this function it fails, since if the chemical yield varies, there is no match between the yields of standard and sample, and all that is achieved is negative - a destruction of the integrity of the standard. his has been discussed in detail elsewhere.[20]

4. SOME RESULTS OBTAINED BY APPLICATION OF RADIOACTIVE TRACERS

Results of RNAA analyses presented in Table 2 were obtained after correction for recovery determined by the use of radioactive tracers. It can be very clearly seen that chemical yields can vary by more than 30 %. This is the case especially when environmental samples with relatively complex matrices are analyzed. The dissolution procedure, which can take several hours, is the main source of losses.[12] Additional uncertainty arises from the separation procedure but this is much easier to control under appropriate experimental conditions. However, in spite of the large yield variations (Table 2), the results after correction for recovery, display good reproducibility and accuracy.

Even when destruction (dissolution) of the sample is performed in a closed system and the separation procedure involves only a small number of steps, the recovery is still not sufficiently reproducible to be taken as a fixed value, nor should it be as a matter of principle since it may easily be measured. Hence, use of radioactive tracers for determining the chemical yield for each analyzed sample is part of the quality control process necessary to obtain high-quality data.[15]

Table 2 *Some results obtained after correction for recovery determined by the use of radioactive tracers.*

Sample	Element	Result Chemical yield		Reference
BCR-CRM No. 146 Sewage Sludge	Pb	mg.g^{-1} 1.24; 1.29; 1.29; 1.22 1.32; 1.23; 1.25; 1.28 ± s.d. = 1.27 ± 0.028 c.v. = 1.27 ± 0.028	% 87; 35; 90; 83 77; 60; 75; 23	12
	Cd	mg.g^{-1} 75.9; 70.7; 72.7; 79.5; 77.0; 77; 74.3; 74.5; 74.4 ± s.d. = 74.8 ± 2.55 c.v. = 77.7 ± 2.6	% 85; 96; 40; 75 81; 83; 93; 84 92	12
NIST SRM 11633a Coal Fly Ash	Cd	mg.g^{-1} 1.06; 1.07; 1.08; 1.06 ± s.d. = 1.06 ± 0.01 c.v. = 1.0 ± 0.15	% 68; 64; 84; 90	12

We also showed recently15 that use of 81mSe for yield determination in RNAA of selenium lead to an improvement in the accuracy and the coefficient of variation of the data.

5. DISCUSSION

RNAA is often considered as a reference technique in certification of reference materials, and/or especially important in special situation when high sensitivity for a particular element in a difficult matrix is required, or no other feasible method exists. The advantage of RNAA incorporating determination of chemical yields should be emphasized as a major feature of this technique. Indeed, it can only be considered a reference method if use is made of this in-built option for quality control. This possibility rests on the isotopic character of NAA; we determine in fact a particular isotope of an element as an activated radioisotope, not the element itself, and this allows the use of carriers and radioisotopic tracers.

Of course there are limitations to the method, such as the lack of suitable radioisotopic tracers in some cases, or for short-lived ones, lack of access to a nuclear facility. Nevertheless, in regulatory work, intercomparisons, referee analyses, RM certification, and in the preparation of protocols for harmonization in the field of recovery factors and trace analysis in general, the claims of RNAA employing chemical yields should be strongly considered.

6. REFERENCES

1.   A. R. Byrne, *Fresenius J. Anal. Chem.*, 1993, **345**, 144.
      A. R. Byrne, Z. Šlejkovec, M. Dermelj, *J. Radioanal. Nucl. Chem., Articles,*1993, **173**, 357.
2.   Eurachem, Quantifying uncertainty in analytical measurement, First Edition, 1995.

3.   K. Heydorn, Neutron Activation Analysis for Clinical Trace Element Research, CRC Press, Boca Raton, FL, 1984.
4.   A. R. Byrne, The preparation and use of chemical irradiation standards, IAEA-TECDOC-323, 1984, 107.
5.   P. De Bievre, *Accreditation and Quality Assurance in Analytical Chemistry*, 1995, **1**, to be published.
6.   Z. B. Alfasi in Chemical analysis by nuclear methods, Edited by Z. B. Alfasi, John Wiley & Sons, Chicester, 1994, 155.
7.   A. R. Byrne, *Biol. Trace Element Res.*, 1994, **43-45**, 529.
8.   H. Schelhorn, M. Geisler, A. Rämmel, *J. Radioanal. Nucl. Chem., Articles*, 1993, **168**, 265.
9.   H. Schelhorn, M. Geisler, *J. Radioanal. Nucl. Chem.*, 1984, **83**, 5.
10.   A. R. Byrne, Vestnik. Slov. Kem. Druš., 1985, **32**, 311.
11.   A. Fajgelj, A. R. Byrne, *J. Radioanal. Nucl. Chem., Articles*, 1995, **189**, 333.
12.   A. R. Byrne, M. Dermelj, *Biol. Trace Element Res.,*1994, **43-45**, 87.
13.   A. R. Byrne, I. Krašovec, *Fres. Z. Anal. Chem.*, 1988, **332**, 666.
14.   V. Stibilij, M. Dermelj, A. R. Byrne, *J. Radioanal. Nucl. Chem., Articles*, 1994, **182**, 317.
15.   H. Schelhorn, M. Geisler, A. Rämmel, *J. Radioanal. Nucl. Chem., Articles*, 1993, **168**, 265.
16.   L. Benedik, A. R. Byrne, *J. Radioanal. Nucl. Chem., Articles*, 1995, **189**, 325.
17.   A. R. Byrne, L. Benedik, *Talanta*, 1988, **35**, 161.
18.   A. R. Byrne, L. Benedik, *Sci. Total Environ.*, 1991, **107**, 143.
19.   A. R. Byrne, *J. Radioanal. Nucl. Chem., Letters*, 1985, **93**, 242.

Measurement Uncertainty: the Key to the Use of Recovery Factors?

Stephen L. R. Ellison[1] and Alex Williams[2]

[1]LABORATORY OF THE GOVERNMENT CHEMIST, QUEENS ROAD, TEDDINGTON, MIDDLESEX TW11 0LY, UK

[2]19 HAMESMOOR WAY, MYTCHETT, CAMBERLEY, SURREY GU16 6JG, UK

1. INTRODUCTION

Evaluating the uncertainty associated with a result is an essential part of any quantitative analysis. In analytical chemistry, perhaps the most reliable method in general use combines extensive interlaboratory comparisons with careful measurement of important performance characteristics of methods to obtain essentially separate estimates of precision and trueness. One such performance characteristic is the 'recovery'. So far, it has not been common practice to include method bias terms, such as method recovery, in reported uncertainties, and recovery may be either reported in addition to confidence intervals or ignored in reporting. Further, the recovery reported may be the typical recovery for a method or, more often, a recovery measured by spiking one sample in a batch to give an indication of acceptable performance. Many users of analytical data have yet to come to terms with two numbers - result and uncertainty - instead of one, to say nothing of a third. Clearly, reporting separate recovery, especially with different meanings, will be undesirable in many instances.

Recently, a standardised approach to the comprehensive determination of uncertainty has been published by ISO[1], and subsequently interpreted for analytical chemistry by Eurachem[2]. The most important characteristic of measurement uncertainty (MU) determination is that MU encompasses random and systematic effects to give a single value or range.

In this paper, we briefly review the definition of measurement uncertainty and its estimation according to the ISO approach. We show how ISO uncertainty estimation can incorporate recovery and its uncertainty, that this is possible irrespective of whether recovery factors are applied, and that, having incorporated recovery information, MU can offer a general answer to the question of how to deal with recovery information in reporting analytical results.

2. MEASUREMENT UNCERTAINTY

There will always be an uncertainty associated with the result of an analysis, for example, because of the random variations that affect the result, uncertainty in factors used to correct for systematic errors, or because of the uncertainty on the validity of the assumptions made about the analytical process. In general there will be many sources of uncertainty and the uncertainty components arising from them must be evaluated and combined in a consistent manner. The general methodology for the evaluation of measurement uncertainty is given in the ISO Guide[1] and amplified for the particular case of chemical measurement in the Eurachem guide[2]. Introductions to the topic and its implications for chemistry can also be found elswhere[3-5]. It is not appropriate to repeat the discussion in this paper but a brief description of both the concept of uncertainty and its evaluation is helpful before discussing its relationship to the use of recovery factors.

Measurement Uncertainty is defined by ISO[1,6] as

"A parameter, associated with the result of a measurement, that characterises the dispersion of the values that could reasonably be attributed to the measurand",

with the note that "The parameter may be, for example, a standard deviation (or a given multiple of it), or the half width of an interval having a stated level of confidence". The ISO Guide recommends that this parameter should be reported as either a **standard uncertainty,** denoted **u**, defined as the

"uncertainty of the result of a measurement expressed as standard deviation"

or as an **expanded uncertainty**, denoted **U**, defined as

"a quantity defining an interval about the result of a measurement that may be expected to encompass a large fraction of the distribution of values that could be attributed to the measurand". The expanded uncertainty is obtained by multiplying the standard uncertainty by a coverage factor, which in practice is typically in the range 2 to 3.

To evaluate the uncertainty systematically it is first necessary to identify the possible sources of uncertainty. Some of the sources relevant in chemical measurement are listed in Table 1 overleaf. The uncertainty arising from each source is then quantified and expressed as a standard deviation, associated with one or more of the intermediate parameters used in calculating the final result. The resulting numerical values, or components, are combined to obtain the overall uncertainty on the basis of their value and the contribution of the parameter affected to the overall result. Again the detailed procedures are given in references 1 and 2. In practice the task is simplified by the predominance of only a few components; others need not be evaluated in detail.

For the present discussion, we are most concerned with the uncertainties covered in items 1 and 3 of table 1, which broadly cover the definition of the measurand (the exact specification of what is to be measured) and effects arising from the sample matrix, including extraction recovery and matrix effects on observed response.

Table 1. Sources of uncertainty in analytical chemistry

1.	Incomplete definition of the measurand (for example, failing to specify the exact form of the analyte being determined).
2.	Sampling - the sample measured may not represent the defined measurand.
3.	Incomplete extraction and/or pre-concentration of the measurand, contamination of the measurement sample, interferences and matrix effects.
4.	Inadequate knowledge of the effects of environmental conditions on the measurement procedure or imperfect measurement of environmental conditions.
5.	Cross contamination or contamination of reagents or blanks[Note 1]
6.	Personal bias in reading analogue instruments.
7.	Uncertainty of weights and volumetric equipment.
8.	Instrument resolution or discrimination threshold.
9.	Values assigned to measurement standards and reference materials.
10.	Values of constants and other parameters obtained from external sources and used in the data reduction algorithm.
11.	Approximations and assumptions incorporated in the measurement method and procedure.
12.	Variations in repeated observations of the measurand under apparently identical conditions.

Note 1. Spurious, one-off contamination would normally represent loss of control of the method, invalidating any prior study of uncertainty. The contamination or cross-contamination referred to here might arise from routine background levels of material under appropriate control, or might constitute a recognition of an unavoidable degree of contamination

2.1 Definition of the measurand

Accurate definition of the measurand is crucial to uncertainty estimation and to the relevance or otherwise of recovery factors. The most important issue is whether the measurand is an estimate of the amount of material actually present in the sample matrix, or the response to a reproducible, but otherwise essentially arbitrary, procedure established for comparative purposes. The latter case is sometimes referred to as an 'empirical method'. This clear distinction between empirical determinations and measurements independent of method makes it essential to define the measurand adequately before considering either recovery or uncertainty. (A secondary issue in measurand definition is the degree to which the analytical procedure itself is specified, but this is beyond the scope of the present discussion).

To illustrate the point, a concrete, or, more accurately, ceramic, example is helpful. Consider the case of lead in ceramic ware. In principle, the measurand might be specified as the amount present - most importantly, *independent of the procedure used*. Alternatively, and in practice, regulation is based on one of a number of well-established procedures for extraction and analysis, such as BS 67487. These procedures are intended to give some idea of the amount extracted under conditions of use; they make no claims on the actual amount of lead present or extractable under other circumstances. It will be clear that where the measurand is specified as 'lead' some knowledge of the method recovery is essential for intercomparison and, indeed, to correct for any bias introduced by imperfect recovery. Under these circumstances, a recovery factor must be estimated and either stated or applied as a correction to the observed value. Conversely, specifying the measurand as 'lead by BS 6748' is a clear indication that knowledge of the recovery for a particular sample or matrix is irrelevant to the purpose (though the recovery measured against the amount *present* is naturally largely responsible for the result obtained). All that is required is the observation resulting from correct application of the stated method. No correction for method recovery is applied unless specified as part of the procedure, and stated uncertainties will not include uncertainties arising from method bias. Similar arguments apply for a very wide range of important analytes, such as 'fat', 'dietary fibre' and so on.

Note that application of a 'standard method' makes does not render knowledge of the recovery and its uncertainty unnecessary. Both are essential if the results might be compared with other methods, for example to establish trends over time. In such a case, the measurand is effectively being redefined, leading to (usually) larger uncertainties and relevance of recovery. In routine testing, however, the lesser uncertainty and 'uncorrected' result will normally be cited.

2.2 Recovery and uncertainty

For the present discussion, we define the Recovery R as the ratio of the observed value c_{obs} obtained from an analytical procedure to a reference value cref. cref will, for example, be given as a reference material certified value, measured by an alternative definitive method or determined as a spike addition. In a perfect analysis, R would be identically 1. In practice, factors such as imperfect extraction or matrix-induced instrument response errors give observations across a range of samples which differ from the ideal response. Because

Table 2: Sources of uncertainty in recovery estimation

1	Repeatability of the recovery experiment
2	Uncertainties in reference material values
3	Uncertainties in added spike quantity
4	Poor representation of native or incurred analyte by the added spike
5	Poor or restricted match between experimental matrix and the full range of sample matrices encountered
6	Effect of analyte/spike level on recovery and imperfect match of spike or reference material analyte level and analyte level in samples.

of this, it is normal practice in validating an analytical method to determine, *inter alia*, an estimated recovery R_m for the method. Typically, this estimate will be made on the basis of comparisons between reference material certified values and values observed using the method, or by spiking studies. In any such experiments, the mean recovery will normally be tested for significant departure from 1. In general terms, such a test will consider the question "is $|R_m -1|$ greater than the uncertainty u_{Rm} in the determination of R_m ?", usually specifying a level of confidence. The uncertainty u_{Rm} will invariably take into account the observed standard deviation of results obtained, and will ideally also include any uncertainties in the reference material values, spike concentrations etc. Table 2 gives some sources of uncertainty in measured recovery. The experimenter then performs a significance test of the form

$|R_m -1|/u_{Rm} > t$: R_m differs significantly from 1   [1a]

$|R_m -1|/u_{Rm} < t$: R_m does not differ significantly from 1   [1b]

where t is a critical value based either on a 'coverage factor' allowing for practical significance or, where the test is entirely statistical, $t_{(\alpha,n-1)}$, being the relevant value of Student's t for a level of confidence $1-\alpha$.

Following such an experiment, four cases can be distinguished, chiefly differentiated by the use made of the recovery R_m:

1. R_m, taking into account u_{Rm}, is not significantly different from 1. No correction is applied.

2. R_m, taking into account u_{Rm}, is significantly different from 1 and a correction for R_m is applied

3. R_m, taking into account u_{Rm}, is significantly different from 1 but, for operational reasons, a correction for Rm is not applied

4.   An empirical method is in use. R_m and u_{Rm} are determined for future information only.

Given that the recovery is neither identically 1 nor exactly known, a complete estimate of uncertainty must take the recovery and its associated uncertainty into account. We now consider how uncertainty is estimated in each of the cases above.

2.3   Contribution of Recovery to Overall Uncertainty

2.3.1 Case 1: R_m *not significantly different from 1*. The experiment has not found any reason to adjust subsequent results for observed recovery. It might be argued that the uncertainty in the recovery is unimportant. However, it should be noted that the experiment could not have distinguished a range of recoveries between $1- k.u_{Rm}$ and $1+k.u_{Rm}$. It follows that the uncertainty about the recovery of the method still applies and should be taken into account in calculating the overall uncertainty. u_{Rm} is therefore included in the calculation. (An alternative argument is that a recovery factor based on $R_m=1$ is implicitly applied, but the experimenter is uncertain that the value is identically 1).

2.3.2 Case 2: R_m *differs from 1 and a correction is applied.* Since R_m is explicitly included in the calculation of the result, it is clear that u_{Rm} will be included (the exact contribution probably being u_{Rm}/R_m, given that the correction is likely to be praltiplicative). This leads to a combined uncertainty utot given by utot2 = (uc/cobs)2 + (u_{Rm}/R_m)2 (using the standard formulae in ref 1), where cobs is the observed result and uc its assigned uncertainty, calculated from all other known components. utot would be multiplied by k (usually 2) to obtain the expanded uncertainty Utot.

2.3.3 Case 3: R_m *differs from 1 and no correction is applied.* Failure to apply a correction for a known systematic effect is not generally consistent with obtaining the best possible estimate of the measurand. It is not surprising, therefore, that it is less straightforward in this case to take recovery and its uncertainty into account in calculating the overall uncertainty. Clearly, if R_m is substantially different from 1, the dispersion of values of the measurand c is unlikely to be properly represented by $u_{tot}^2 = (u_c/c_{obs})^2 + (u_{Rm}/R_m)^2$ unless the uncertainty u_{Rm} is substantially increased. A simple and pragmatic approach that is sometimes adopted when a correction **b** for a known systematic effect has not been applied is to increase the expanded uncertainty on the final result to **(U+b)** where U is the expanded uncertainty calculated assuming **b** is zero. For recovery, therefore, $U_{tot} = U_c + c_{obs}/R_m$. This procedure gives a somewhat pessimistic overall uncertainty, and moreover departs from the ISO-recommended principle of treating all uncertainties as standard deviations.

Alternatively, if the correction for recovery is not applied because it is felt that though the statistical test (as at case 2) shows a difference, the analyst's judgement is that the difference is not practically significant in normal use, case 3 may be treated in the same way as case 1 after increasing u_{Rm} to account for the analyst's implicit recognition that the significance test should have used a larger value than u_{Rm}. This amounts to estimating u_{Rm} as $|1-R_m|/t$, where t is the critical value used in the significance test at 1a and 1b. (2 for approximately 95% confidence on a large sample).

In addition to increasing the stated uncertainty to allow for imperfect recovery, the uncertainty on recovery should be included as usual. This will normally only be significant where u_{Rm} is comparable with or greater than $|1-R_m|$.

Note that while either method will provide a viable estimate of uncertainty, both methods have similar drawbacks arising directly from the failure to correct the result to give a best estimate of the value. Both lead to some overstatement of the uncertainty, and in both cases, the range quoted around the result will include the most probable value of the measurand only near one extreme (usually the upper end), with the remainder of the range unlikely to contain the value with significant probability. Alternative approaches are available for such situations, including estimating separate uncertainties for the range above and below c, or allowing explicitly for asymmetric distribution functions in calculating the appropriate range, but only at increased cost in formulating the estimate. In the view of the present authors, these difficulties form a case either for eliminating the recovery error by altering the method or for applying the correction, rather than any argument against increasing the uncertainty estimate.

2.3.4 Case 4: Use of an empirical method. In this case it is not the amount of analyte in the sample that is being determined but the result obtained by applying the procedures set out in the empirical method. This should be made clear when reporting the result. Whether or not a correction is applied for recovery depends on what is specified in the empirical method. The uncertainty on the result depends chiefly upon the performance characteristics of the method and on how well the procedures set out in the method have been followed, with some additional allowance for items not completely specified (for example, not all methods specify the method of instrument calibration exactly). In estimating overall uncertainty, the uncertainty on the recovery should be included if a measured recovery is used to correct the result, but if the method specifies a fixed recovery factor no additional uncertainty term arises. The problems of evaluating the uncertainty in this case are discussed elsewhere[3].

To compare the result obtained with an empirical method with that obtained by a method which attempts to determine the actual amount of the analyte in the sample it is necessary to determine the correction for the bias of the empirical method when utilised on that particular sample as well as the uncertainty on this correction.

2.4 Implications of estimating uncertainty with imperfect recovery

The cases above show that where recovery is an issue, a comprehensive estimate of uncertainty can be obtained by applying ISO methods, whether recovery is found to be essentially complete, significantly different from unity and corrected for, or uncorrected, though the latter case requires as much pragmatism in the uncertainty estimate as is implied by treating the correction as insignificant. It is worth considering briefly how these approaches will affect the uncertainty estimate.

In case 1, we have asserted that the recovery, taken as unity, nonetheless has some uncertainty, leading to a moderate increase in overall uncertainty. Case 2 results in a similar situation; including the recovery uncertainty is good practice, and leads to moderate increase in overall uncertainty. Case 3, on the other hand, may add a very significant amount to the uncertainty estimate. For recoveries of the order of 70%, the additional uncertainty contribution (before applying a coverage factor) will be close to 20% of the result. This is clearly not unreasonable given the size of recovery correction being ignored, but it does point strongly to the consequences for reported uncertainty of neglecting a substantial recovery correction; in the case of a 70% recovery, the reported uncertainty after application of a coverage factor will exceed 40% of the result.

There is therefore a clear choice if the customer is not to be misled by a result and associated uncertainty. Either the recovery must be corrected for, typically by application of a recovery factor, or a substantially greater uncertainty must be quoted.

Finally, it should be noted that the foregoing discussion relates to the situation where result and uncertainty are reported to the eventual end user. We have not considered the case where an analyst provides an interpretation of a result, for example in the form of a report stating that the value is "Not less than...". It is taken for granted that in this kind of interpretation, the analyst's professional knowledge of the recovery and overall experimental uncertainty will be taken into account in the interpretation, and accordingly neither the recovery nor an uncertainty need necessarily be reported

3. CONCLUSIONS

Estimated measurement uncertainty should normally incorporate an allowance for uncertainty in estimated recovery where the measurand in principle requires a correction for recovery. In this paper, we have presented methods of allowing for recovery and its uncertainty in calculating overall uncertainty, whether or not the correction is applied. It is shown that the failure to apply a correction for a known effect will lead to substantially greater uncertainty estimates, pointing to correction for recovery as providing a smaller overall uncertainty.

The chief exception to this general principle is the case of empirical methods, in which the method defines the measurand. Whether or not a correction is made for recovery depends upon what is specified in the method, and will not affect the use of the method for comparative purposes. However, if the result is to be compared with that obtained from another method, the recovery factor, any other bias of the method and their uncertainties will need to be taken into account.

4. ACKNOWLEDGEMENT

The authors acknowledge the support of the Department of Trade and Industry through the Valid Analytical Measurement initiative.

5. REFERENCES

1   'Guide to the Expression of Uncertainty in Measurement', ISO, Geneva, 1993, (ISBN 92-67-10188-9)

2   'Quantifying Uncertainty in Chemical Measurement', Laboratory of the Government Chemist, London 1995. ISBN 0-948926-08-2.

3   Analytical Methods Committee, *Analyst*, 1995, **120**, 2303

4   Williams, A., *Accreditation and Quality Assurance*, 1996, **1**, 1

5   Cortez, L., Mikrochimica Acta, 1995, **119**, 323

6   *International Vocabulary of basic and general standard terms in Metrology*. ISO, Geneva, Switzerland 1993 (ISBN 92-67-10175-1)

7   British Standard BS 6748: 1986. Limits of metal release from ceramic ware, glassware, glass ceramic ware and vitreous enamel ware.

Varying Uses of Recovery Factors in US EPA FIFRA Registration Studies

Del A. Koch, Patrick A. Noland, and Loren C. Schrier

ABC LABORATORIES, 7200 E. ABC LANE, COLUMBIA, MO 65202, USA

1. INTRODUCTION

ABC Laboratories is a contract research organization providing services which include the performance of studies to support the registration of pesticides. These studies are required by the United States Environmental Protection Agency (US EPA) as described in the Federal Insecticide, Fungicide, and Rodenticide Act (FIFRA). Recovery factors may be used in various ways for these studies, depending upon the type of study and the preferences of the registrant. The information presented herein is based upon typical practices as well as guidance documents provided by the US EPA and the Food and Drug Administration (FDA). However, the authors wish to stress that the following discussion does not necessarily represent official EPA or FDA policies.

2. MAGNITUDE OF THE RESIDUE STUDIES

Method recoveries in magnitude of the residue (MOR) studies are primarily used for evaluation of the acceptability of sample analysis data. With each set, or batch, of samples analyzed concurrently, at least one control sample fortified with the analyte(s) is analyzed for quality control purposes. Typically, the number of fortified samples analyzed with each set will represent 10-20% of the total number of samples in the set. Typically, recoveries in the range of 70-120% are considered to be acceptable and thus validate the results of the other analyses within the same set. In some instances, a wider range of acceptability may apply for fortifications performed at or near the limit of quantitation. For methods with an inherent low bias, lower recoveries (60-70%, for example), may be acceptable if it is demonstrated that recoveries are consistently within a relatively narrow range.

US EPA allows for recovery factors to be used in the correction of residue values generated in magnitude of the residue studies, but does not require that this correction be made.[1,2]

Typical practice is that no correction is made to residue levels determined in MOR studies. These studies are designed to determine the highest residues likely to be observed so that tolerances may be set, typically at a level above the highest observed levels. These tolerances are generally well below any level of toxicological concern, but they determine the maximum allowable levels for enforcement purposes. US FDA monitors the United States food supply and may utilize methodology supplied by the registrant. If the same practice (i.e., not correcting for recovery) is used both for setting and enforcing the tolerance level, the data should be consistent. If correction factors <u>are</u> used, it must be specified whether the correction factor is based upon an average recovery (i.e., the average of all recoveries generated from several sets of sample analyses) or upon the recovery from each set.[2] Each technique has some possible limitations. While a recovery from a given set may be a statistical outlier and not representative of method performance for the samples, the overall method recovery will not address the issue of day-to-day variability in the method performance. The optimum practice would be to perform several recoveries with each analysis set to obtain a statistically valid average for that specific set, but this may not always be practical or cost-effective.

Additionally, if correction factors are used, the uncorrected residue values must be reported along with the corrected residue values.

3. STORAGE STABILITY STUDIES

US EPA requires freezer storage stability data in support of studies in which samples are stored frozen for a period of greater than 30 days between sampling and analysis. These studies involve fortification of sample matrix with the analyte(s), storage of the fortified samples under freezer conditions, and analysis of the stored samples at specified timepoints. For these studies, US EPA specifies that freshly fortified samples be analyzed at each timepoint and that the results of the fresh fortifications be used to correct the results of the freezer stability sample analyses.[3] Typical practice is to perform single or duplicate fresh recoveries at each analysis timepoint. The single fresh recovery or the average of the duplicate fresh recoveries would then be used to correct each storage stability sample from that given timepoint.

4. US FDA MULTIRESIDUE METHODS TESTING

One of the US EPA testing requirements is that a pesticide must be tested through the US FDA multiresidue testing methods. The information generated by this testing will go into the FDA database ('PESTDATA'). In this testing, correction for method recoveries is not relevant, since the purpose of the testing is to determine recoveries of the analyte(s) through the various methods, not to analyze samples. However, the recovery ranges are classified somewhat differently that the acceptance criteria specified by the US EPA for residue analyses, and thus warrant mention.

For some of the multiresidue methods, the testing of analyte(s) recovery from Florisil (an adsorbent used in a cleanup column step) is a prequalification step performed prior to analyzing fortified sample matrix through the entire method. In this case, recovery of 30% or greater is sufficient for proceeding with the testing. Recoveries through the various full multiresidue methods are classified as: >80%, complete; 50-80%, partial; <50%, small; and not recovered.[4]

5. ENVIRONMENTAL FATE STUDIES

Terrestrial (soil), aquatic, and forestry dissipation studies, as well as groundwater studies, are required under the Environmental Fate (Subdivision N) guidelines. The use of recovery data for samples analyzed in conjunction with these studies is similar to that described previously for the magnitude of the residue studies. Recovery data are primarily used for quality control purposes, with recoveries in the range of 70-120% indicating acceptability of the sample analyses. In general, corrections for recovery may be made, but this is not required.[5] One exception is (for soil dissipation studies) the confirmation of the reported application rate by analysis of soil samples taken immediately after treatment. Typically, the residues found in these soils must correspond to at least 70% of the label application rate. For this determination, correction for method recovery is not permitted.

It is our experience that the correction of soil analysis residue results for method recovery is a rarity. Much more common, for example, is correcting for moisture content and presenting the data on a dry weight basis, thus normalizing the results generated from different sampling intervals.

6. FORMULATION ANALYSIS

As part of the US EPA product chemistry registration requirements, formulation analysis methods are required for the quantitation of each active ingredient and each impurity for which a certified limit is required (with certain exceptions). US EPA specifies precision requirements based upon the concentration of the analyte in the formulation. This information is reproduced as Table 1.[6]

Table 1   *Recommended Levels of Precision for Formulation Analysis*

Ingredient Measured in Sample	Precision: 100 X (standard deviation , arithmetic mean)
More than 10.0%	Not greater than 2%
1.0% up to 10.0%	Not greater than 5%
0.1% up to 1.0%	Not greater than 10%
Less than 0.1%	Not greater than 20%

Typically these formulation analyses do not require extensive cleanup and therefore the low bias inherent in many residue levels methods may not be present. Thus, for example, recoveries for components present at a level of 10% or greater should be recovered within the range of 98-102%, components present at a level between 1.0 and 10.0% should be recovered within the range of 95-105%, and so forth.

Analyses using previously approved and validated methods may not require concurrent recovery analyses. Typically, formulation analysis results are not corrected for method recovery. One interesting exception to this practice has been presented previously by our laboratory.[7]

ABC Laboratories was contracted to develop an assay for an active ingredient that was a water soluble sulfate salt of a multicomponent aminoglycoside antibiotic. Two formulations were tested, one at an active ingredient level of 1% and one at 10%. Both formulations were wettable powders and consisted of the active ingredient in an "inert" clay matrix. The initial method development extractions with pH neutral H_2O yielded 0% recovery from the 1% formulation and less than 80% from the 10% formulation. After various investigations it was determined that a 1:1 mixture (v/v) of 0.1 M $Na_2B_2O_7$ (pH 11.5-12.0) and methanol yielded recoveries that were consistently in the 70-80% range for the 1% formulation and in the 90-100% range for the 10% formulation. Five recoveries at the 1.0% level yielded an average recovery of 76% with a standard deviation of 3.5%, for a precision of 4.6%. The method thus met the precision requirements, but the use of a correction factor was necessary because of the low bias. The method as written thus specifies that results obtained by the method on the 1% formulation should be divided by the correction factor of 0.76 unless recovery tests are performed along with the analysis of the formulation samples.

7. CONCLUSIONS

For US EPA pesticide registration studies performed under FIFRA guidelines, concurrent recovery analyses are performed along with residue-level analyses for quality control purposes to evaluate data acceptability. Although the US EPA guidelines generally allow sample values to be corrected for method recovery, this is not typical practice. One exception is the analysis of freezer stability samples, for which US EPA specifies correction for the concurrent fresh recovery value. For formulation-level analyses, methods normally do not exhibit a significant bias and therefore correction for method recovery is not necessary. However, as outlined previously, there are special cases where significant method bias can make correction desirable.

9. REFERENCES

1. M. J. Nelson and F. D. Griffith, Jr., 'Addendum 4 on Data Reporting to Pesticide Assessment Guidelines, Magnitude of the Residue: Processed Food/Feed Study', US EPA, Accession No. P88-117270, November, 1987.
2. D. Edwards and E. Zager, 'Guidance on Submission of Raw Data', US EPA, January, 1993.
3. D. Edwards and E. Zager, 'Guidance on Generating Storage Stability Data in Support of Pesticide Residue Chemistry Studies', US EPA, January, 1993.
4. US FDA Pesticide Analytical Manual, Volume I, Appendix I, January, 1994.
5. 'Pesticide Reregistration Rejection Rate Analysis Environmental Fate, Follow-up Guidance for: Submission of Raw Data', US EPA, April 1995.
6. 'Pesticide Assessment Guidelines, Subdivision D, Product Chemistry', US EPA, Accession No. PB83-153890, October, 1992.
7. L. C. Schrier, P. A. Noland, and D. A. Koch, 'Inert Carrier Effects and Interactions in Active Ingredient Analysis', presented at the 109th AOAC International Meeting, Nashville, Tennessee, USA, September 20, 1995.

Use of Recovery Data in Veterinary Drug Residue Analysis

James D. MacNeil, Valerie K. Martz, and Joe O.K. Boison

HEALTH OF ANIMALS LABORATORY, AGRICULTURE & AGRI-FOOD CANADA, 116 VETERINARY ROAD, SASKATOON, SASKATCHEWAN S7N 2R3, CANADA

1. INTRODUCTION

There has been no standard approach among residue analysts, in the veterinary drug residue area, with respect to the use of recovery data to correct the results of chemical analyses obtained on analytes extracted or isolated from biological matrices. Recovery, expressed as a percent, is the proportion of the analyte present in a sample matrix that is capable of being determined by an analytical detection system. It can also be defined as the sum of individual losses in analyte concentration obtained in every step of the sample preparation procedure from sampling to the preparation of the final test solution needed for qualitative and/or quantitative analysis. If we consider an analytical method consisting of the four basic steps: sample pretreatment prior to extraction, sample extraction, evaporation to dryness, and purification, and if Q_A is the amount of sample initially present in the sample, and Q_B is the amount of sample that was found after the sample has been treated according to the method, then the overall recovery, R_o, is given by

$$R_o \quad = \quad [Q_B/Q_A] * 100$$

The overall recovery R_o is dependent on the quality of recovery obtained at every step included in the sample preparation procedure beginning from sample pretreatment up to the final sample solution injected into the GC or HPLC column. Recovery at each step R_p, R_e, R_{ev}, and R_{pur}, must therefore be carefully investigated from this respect, since the final results of overall recovery should not be better than the result of partial recovery obtained in the weakest step of the whole procedure. The ability to determine the overall recovery of the analytical procedure for the analyte(s) of interest then hinges very significantly on the ability of the analyst or the laboratory to design proper experiments (recovery experiments) to identify the weakest step in the analytical procedure and to measure the recovery for this step. The overall recovery factor, R_o, if lower than 100%, should be used to correct the analytical result obtained with the analytical procedure.

Standard practice for some analysts, laboratories or government departments is to report uncorrected analytical results, some report uncorrected results with a recovery estimate, while others report recovery corrected values. This can lead to problems in comparison of data sets, or in the comparison of results from different laboratories, whether it is to settle a dispute as to the actual concentration of a residue in a commodity under investigation, or whether the comparison is to determine method or laboratory performance on a check sample. In addition, there has been no standardized approach to the calculation of recovery factors. Thus, two laboratories may report recovery-corrected results on a sample, but may base the recovery correction on different experiments or approaches. The 45th Meeting of the Joint FAO/WHO Expert Committee on Food Additives, meeting in Geneva in June, 1995, to consider residues of veterinary drugs, has requested that in future, all residue dossiers submitted for review should report residue results that have been corrected for recovery to ensure that all results are considered on an equivalent basis (1). Some of the problems faced by the analyst and the approaches taken to estimate the analyte recovery are the topics of this paper.

2. THE USE OF THE RESULT

Analytical results produced by a veterinary drug regulatory laboratory are generally used for one of two purposes. Survey samples are collected randomly to determine the incidence of violative residues for a particular analyte, geographical area and time period. Analytical results do not usually result in an action on the sampled product, which has already moved on in normal commerce, but a positive will result may generate follow-up action on a producer or on subsequent lots of product from a particular source (eg., a slaughter plant or geographic area, when the individual producer is not known). Suspect samples are tested to determine compliance of that specific sample with a residue limit, with regulatory action following on a specific carcass, container of milk or a shipment of a product. These samples have usually been designated as suspect by the use of a screening test in the field and the product is in detention pending receipt of the laboratory result. In survey or suspect sample testing, the analysis is to determine product conformance with a Maximum Residue Limit (MRL), or tolerance, which is based on the Acceptable Daily Intake (ADI), calculated based on the toxicity of the residue. Because the estimate of toxicity is based on the total quantity of residue present, and not the amount of residue that may be extracted by any particular analytical method, it is important to make corrections when using analytical methods that are only capable of partially extracting the residue of interest from the applicable matrix. The residue to be determined is usually designated as a marker residue in the form of either the parent compound or its metabolite, which is representative of the total residue of concern.

Therefore, in determining whether a veterinary product meets the limits that have been set as safe for human consumption, the analyst should ensure that the value reported reflects the best estimate of the total residue of the marker residue that is present.

Let us consider an example. The Codex Committee on Residues of Veterinary Drugs in Foods has recognized an MRL of 0.1 mg/kg for residues of sulfadimidine (sulfamethazine) in pork liver and muscle, a value that has also been adopted in the regulations of many countries (2). Let us now assume that the method in use in a particular laboratory provides an average of 75% recovery of the marker residue. A sample analyzed by the laboratory is found to contain

0.09 mg/kg of sulfadimidine, on an uncorrected basis. Is this sample to be treated as violative? Applying a recovery correction, the estimated sulfadimidine content becomes 0.12 mg/kg, clearly in violation. However, if the corrected result is not reported, or if the regulatory official who must act on the result is not advised that the result is based on 75% recovery, the sample will probably be considered non-violative. Imagine the confusion that can result if different laboratories in different parts of a country or trading bloc use methods that have different recoveries and have different policies on recovery correction. This can lead to an inaccurate picture of the frequency of violations in different areas, simply by the choice of analyzing laboratory. It can also lead to problems in trade, where samples tested and found acceptable in the exporter's laboratory are reported violative by the importer. Clearly, it is desirable that a uniform approach to the use of recovery correction of analytical results be adopted, particularly where the safety of food is the consideration.

3. BASIS OF THE CORRECTION

There is no standard approach to determining the recovery correction used in veterinary drug residue laboratories. Methods are characterized in terms of their accuracy and precision (3, 4). Ruggedness testing is done to determine critical control points. Inter-analyst and interlaboratory testing is done to better characterize the method performance characteristics. We need to bear in mind, however, particularly those of us dealing in trace quantities of organic residues which are extracted from the complex of organic chemicals found in a typical biological tissue sample, that what we are obtaining with our analysis is not an absolute value determined from an intrinsic property of the matter. We are making our best estimate and there are many factors that can influence the validity of that estimate. One of these is analyte recovery and the way in which that is estimated.

3.1 Use of Radiolabelled Analyte

The best estimate of recovery is obtained when a means is provided not only to measure the amount of analyte recovered by extraction, but also to measure what remains in the sample following extraction. Total residue studies conducted to determine the fate of a veterinary drug in an animal usually will include studies with the radiolabelled drug to follow the residues through the metabolic pathways and determine the persistence of residues in various tissues. The radiolabel permits a complete accounting of the total residue. However, the problem is that, while we can further analyze and characterize extracted residues, we cannot do the same for the so-called bound or unextractable residues. We do not know in many cases if they represent potentially active material. The usual approach in a toxicological evaluation is to make a worst case assumption that they are potentially toxic and to take this into consideration in setting the ADI and the MRL. However, this does not account for further loss of analyte which may occur during subsequent analytical steps, such as solid phase extraction, solvent partitioning, evaporation and derivatization.

These losses can be monitored with a radiolabelled residue. The use of such materials, however, is expensive, creates safety concerns and raises waste disposal costs. For these reasons, as well as simple availability, use of radiolabelled analytes spiked into samples on a routine basis is not a common approach to monitor recovery.

3.2 Other Isotope Methods

Isotope labelling techniques are particularly popular in neutron activation analysis and in some mass spectral assays, where an isotope-labelled form of the analyte is added to the sample as an internal standard. It is assumed that under the prevailing experimental conditions, the labelled analyte will exhibit the same behaviour through all steps of the analysis. The mass spectrometer, for example, can discriminate between the ions resulting from the incurred and the added isotope-labelled material, providing data that enable the analyst to estimate the amount of incurred residue present from the abundance of the isotope-labelled ions. The quantity of incurred residue is therefore estimated based on the response relative to that of the added internal standard and, assuming the same recovery of both, the result is automatically recovery corrected. This approach is taken in a commonly used assay for residues of diethylstilbestrol (DES), where deuterium-labelled DES is added as the internal standard (5).

3.3 Non-labelled Internal Standards

A similar approach to isotope labelling, and one that is more commonly used, is to choose a chemical that is structurally very similar to the analyte and to add that as the internal standard. Again, the assumption is that the two compounds will behave virtually identically during the various steps of the analysis, so that a direct comparison may be made between the analytical response for the added internal standard and the incurred residue. The method referenced for DES also is used for the determination of residues of the growth promotant zeranol, but with the addition of the compound zearalane as the internal standard (5). Usually, the compound selected for such use is a metabolite, an isomer or another member of the same class of compounds as the target analyte. Experiments must be conducted to ensure that the compound selected does, in fact, mirror the behaviour of the target analyte. Any deviation can lead to significant errors. This approach is probably the one most commonly used in analytical laboratories when a suitable surrogate is available.

3.4 Analyte Addition

This is a variation on the technique of standard addition. Typically, a known quantity of analyte is added to a blank tissue and then is taken through the analysis, from which the "recovery" may then be calculated. This technique relies on the same assumption as does the internal standard approach, that the behaviour of the added analyte accurately reflects the behaviour of the incurred residue. The weakness in this is, of course, that the incurred residue has passed through the living animal and has been present in tissues and exposed to various enzymes, proteins, etc., for some period of time following drug administration, while the added standard is post-mortem, usually to a sample that has been frozen and thawed. The manner of addition, whether it takes place before or after sample homogenization, and contact time prior to extraction may all influence the recovery of the added material. Generally, this procedure tends to give optimistically high recoveries, because the added analyte does not usually bind to the matrix constituents in the same way that the naturally occurring analyte may be, and so will be recovered more efficiently (6).

The addition of known quantities of analyte to estimate recovery is a common practice. It has the advantage that it can be used easily to monitor recovery at various steps in an assay.

However, it is important to remember that such experiments should be carried out not only with standard solutions, but also in tissue extracts, as co-extractives may influence the results. Once a recovery performance record has been established for analyte spiked into blank sample matrix, one or more spiked samples can then be included with each analytical run to check recovery and these recoveries can be plotted on an on-going basis using control charts as a quality control measure. It is also common practice in many laboratories to use spiked samples to plot a calibration curve for the analysis, instead of using chemical standards taken through the assay for this purpose. If chemical standards are used to plot the curve, it is then important that spiked samples be included in the run as a means of checking recovery. If spiked samples are used to generate the calibration curve, that curve is, in effect, recovery corrected. Again, it must be emphasized that the approach assumes equivalent recoveries for the added and the incurred analytes (7). While this may not always be a valid assumption, it remains the method of choice for many laboratories.

3.5 External Standard

This has been addressed above, where chemical standards are used to produce the calibration curve. This method probably provides the poorest estimate of recovery, unless fortified samples are analyzed to provide a recovery estimate.

4. CONCLUSIONS

At this point, we have probably raised more questions than we have answered, as there remain some questions, such as the key issue of the comparability between the actual recovery of incurred and added residues. This is probably variable, depending on the analyte-matrix combination, and is not readily available with current methods. The problem then becomes to standardize the approach, so that at least results from different laboratories should be more comparable. As a practical measure, we need to define the experiments that should be conducted when we use any of the approaches to the estimation of recovery, or the correction for recovery, that have been discussed above. For those laboratories that use addition of an analytical standard to a blank sample matrix to generate a calibration curve or to provide an estimate of recovery, it is particularly important that the method of addition, the state of the sample when the spiking occurs and the time and conditions of the contact between analyte and matrix prior to extraction be defined. From our perspective, for our type of regulatory work, the question is not should we correct for recovery, but rather the manner in which the recovery correction should be determined.

5. REFERENCES

1. *Evaluation of Certain Veterinary Drug Residues in Food*, 45th Report of the JECFA, WHO Technical Report Series, World Health Organization, Geneva (in press).
2. *Codex Alimentarius, Volume 3, Residues of Veterinary Drugs in Foods, 2nd Edition*. 1993. Food & Agriculture Organization of the United Nations, Rome, Italy.
3. Pocklington, W. D. 1990, *Pure and Applied Chem.*, 62: 149-162.
4. King, B. 1992, *Analytical Proceedings*, 29:184-186.
5. Covey, T.R., Silvestre, D., Hoffman, M.K. and Henion, J.D. 1988, *Biomedical and Environmental Mass Spectrom.*, 15: 45-56.
6. Thompson, M. 1992, *Analytical Proceedings*, 29:190-193.
7. Dabecka, R. W., and Hayward, S. 1993, "Missing Aspects in Quality Control" in *Quality Assurance for Analytical Laboratories*, ed., M. Parkany, R. S. C.

Development of Recovery Factors for Analysis of Organochlorine Pesticides in Water Samples by Using a Micro-Extraction Technique

R. Boonyatumanond, M.S. Tabucanon, P. Prinyatanakun, and A. Jaksakul

DEPARTMENT OF ENVIRONMENTAL QUALITY PROMOTION, ENVIRONMENTAL RESEARCH AND TRAINING CENTER, TECHNOPOLIS KLONG LUANG DISTRICT, PATHUMTHANI 12120, THAILAND

Abstract

Water samples have been analyzed for 22 compounds of organochlorine pesticides by using micro-extraction technique. The technique was easy to perform, flexible and required minimal glasswares and sample handling. This approach not only fulfilled the practical requirements of this program but also gave reliable data. A method based on simple extraction theory for determining recoveries and quality control by using oxychlordane and pentachloronitrobenzene (PCNB) as internal standards. A step of recovery and recovery factor were considered to compensate extracting and transferring of analyte from the matrix. Recovery factor of oxychlordane was used for correction volume of extracting solvent. HP-1 (crosslinked to methyl silicone gum) 0.2 mm. id x 50 m x 0.5 μm film thickness was used to separate 22 compounds in 45 min. The detection limit was 0.02 - 0.10 ppb and recovery 75-90 % which depend on chemical properties of each organochlorine pesticides .

1. INTRODUCTION

Organochlorine pesticides were worldwidely used to increase agricultural yield and protect livestock. Environmental and human health hazards by the use of these chemical in Developing Countries are discussed. On the other hand , organochlorine pesticides are toxic and can be hazardous to human being because of bioaccumulation , persistence in the environment[4] and food chain[7] such as HCHs, aldrin, dieldrin, endrin and DDTs etc. Development of analytical method for determination 22 compounds of organochlorine pesticides in water sample by using microextraction technique was introduced to rapid screening in river water sample.

The main objective of this development program is to determine the organochlorine pesticide residues within the shortest possible time , small amount of solvent or chemical and required minimal glassware.

Microextraction technique employed common technique perform by using small amount of solvent for extraction. The method based on simple extraction theory for determining recoveries with spikes in water sample. In some cases an internal standard was added to the extracting solvent to compensate for injection error and solvent evaporation[2,7]. Because this technique used a volatile organic solvent to extract chlorinated compound from water sample.The analysis of water samples were extracted with volatile organic solvent[3], in which the solvent-to-water ratio ranged from 1:40 to 1:1000[7]. This approach was developed for use in the U.S.Environmental Protection Agency (EPA) verification sampling and analysis program.

Recovery factor was considered for quality control in quantitative data for each step.The 3 recovery factors investigated matrix ,spike level for determination detection limit and optimum of extraction time .

2. EXPERIMENT

2.1 Apparatus

(a)Extraction system . Mechanical rotation model SR 550 (Advantec)

(b)Gas chromatograph .Hewlett packard 5890 series II plus ,equipped with a Ni63 electron capture detector . HP-1(crosslinked methyl silicone gum) capillary column (50m x 0.2 mm.id x 0.5 μm film thickness) was used to determine the 22 target compounds. The temperature program was as follows: The initial column temperature was held at 70 °C for 1 min. after injection and increased at 20°C/min to 150°C hold 1min.and increase 3°C/min to 185°C hold 1min. and increase 1°C/min to 198°C hold 18 min. The injection and detector were 220 °C and 300 °C respectively. The carrier gas is hydrogen (flow rate ca 0.7 ml/min). The nitrogen was used for make up gas (flow rate ca 40 ml/min) .

(c)Vortex genic 2. model G-560 E (USA)

2.2 Material

(a) Standards. 22 target compounds were obtained from GL science Inc,Japan.The purity of all compounds were greater than 98%. Stock solution of the 22 target compounds were prepared in pesticide grade hexane 200 μg/ml ; working solution were prepared by using serial solution of composite stock solution about 10 ng/ml .

(b) Solvent . Hexane , acetone (Pesticide grade from J.T.Beaker,USA)

(c) Internal standard . Oxychlordane and pentachloronitrobenzene (PCNB) were obtained from Wako Pure Chemical Industrials LTD., Japan.

2.3 Method

Water sample 1 liter was extracted with 2 ml hexane by using mechanical rotation for 15 ,30 ,60 and 120 min,respectively and stand for seperation. The extract was transfered to a new tube and add 0.5 ml oxychlordane, mixed well and was transfered 1 ml to a new tube and add 0.5ml PCNB and injected into GC. The clean up of extracted sample may be include if needed.

3. EXPERIMENTAL DESIGN

The experimental design consisted of the experiment runs as shown in Table 1 . Each time period were replicated 4 times to determine the optimum extraction time for microextraction technique. 2 matrix types were used (trap water and river water) Concentration for recovery test 3 level were used the mixed standard in river water to determine the detection limit at 12 replication in each levels as shown in Table 2.

In this study recovery factor (RF) were used to correct the extraction time and detection limit by using internal standards (oxychlordane and pentachloronitrobenzene).

Table 1 Experimental design

A. The matrices were checked for desicion further clean up step by analysing tap water sample and river sample .

B. Determination of the extraction time

Chemical	Extraction time (min)			
Standard mixture	15	30	60	120

C. Concentration of spiked mixed standard for determination of method detection limit.

Chemical	conc^n of spike	concentration after spiking (ng/l)		
		0.5ml	0.3ml	0.2ml
α-HCH	31	15	9.2	6.1
HCB	41	20	12	8.2
β-HCH	316	158	95	63
γ-HCH	63	32	19	13
δ-HCH	100	50	30	20
Heptachlor	104	52	31	21
Aldrin	63	20	19	13
Heptachlor epoxide	58	50	17	12
oxy-chlordane	40	20	12	8.0
trans-chlordane	60	30	18	12
o,p-DDE	135	67	40	27
Endosulfan I	417	208	125	84
cis-chlordane	104	52	31	21
Dieldrin	90	45	27	18
p,p'-DDE	100	50	30	20
o,p-DDD	150	75	45	30
Endrin	126	63	39	25
Endosulfan II	158	79	47	32
p,p'-DDD	125	63	38	25
o,p-DDT	158	79	47	3
Endosulfan sulfate	245	122	73	49
p,p'-DDT	274	137	82	55
Methoxychlor	405	202	121	81

4. CALCULATION

(1) The recovery factor (X1) was used to determine extraction volume by using oxychlordane as internal standard .

$$\frac{\text{spiked concentration} \cdot \text{spiked volume}}{X_1} = \frac{\text{final volume} \cdot A \cdot C}{B}$$

where X_1 = extraction volume (ml.)
   A = peak area of oxychlordane in sample
   B = peak area of oxychlordane in standard
   C = concentration of oxychlordane (ng./ml.)

(2) Determination of recovery factor X2 by using PCNB as internal standard in certain technique to determine injection error.

5. RESULTS AND DISCUSSION

The results shown that this method can analysed the 22 target compounds that were spiked in water samples .The river water and tap water sample were tested to skip the clean up step . The chromatogram of river water samples was very clear to identify the target compounds in Figure1, that were compared with a chromatogram of river water sample by ordinary method as shown in Figure 2 . and also the chromatogram of spiked mixed standard in water sample was clear in Figure 4.when cmpared with a chromatogram of standard in Figure 3.

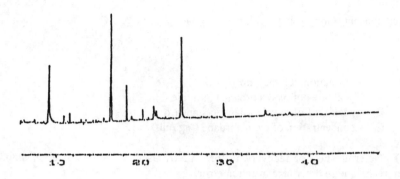

Figure 1   Chromatogram of river water sample
            by using microextraction technique

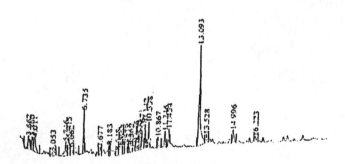

Figure 2   Chromatogram of river water sample
            by using ordinary method

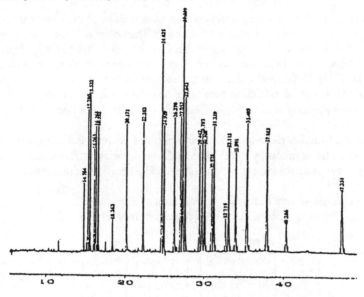

Figure 3 Chromatogram of 24 mixed standards

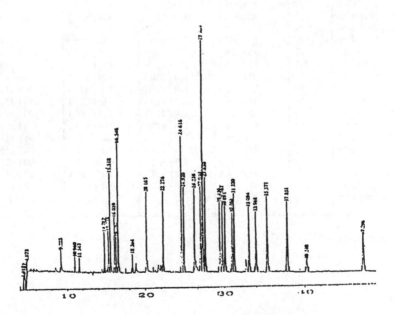

Figure 4 Chromatogram of spiked river water sample

Determination of the optimum extraction time were confirmed by 4 replication of each time period (15 min, 30 min, 60 min. and 120 min). The optimum time was 60 min. and appropriate to design all the experiment because of small %RSD and high of the percentage of recoveries by using recovery factor as shown in Table 3 and ordinary calculation in Table 2. The method of using recovery factor shown that variation of a recovery and %RSD of all extraction time were less than ordinary calculation, especially %RSD of 60 min the level was <10% , for example α -HCH, HCB and β-HCH as shown in Figure 5 and 6.

In the microextraction technique, the evaporation of solvent and the recovery factor was considered as factor necessary to calculate as shown in formula. The recovery factor X1 and X_2 were used to determine extraction volume and error of injection into the GC

Table 2. Experimental design for optimization of extraction
by using general calculation of percent recovery.

Chemical	Spike (ng/L)	%R.(±%RSD) 15 mins	%R.(±%RSD) 30 mins	%R.(±%RSD) 60 mins	%R.(±%RSD) 120 mins
α-HCH	15	75 (±20)	100 (±14)	94 (±30)	89 (±17)
HCB	20	48 (±25)	52 (±14)	53 (±32)	45 (±19)
β-HCH	160	87 (±22)	130 (±16)	110 (±30)	100 (±23)
γ-HCH	32	75 (±22)	99 (±15)	94 (±30)	82 (±19)
δ-HCH	50	55 (±44)	63 (±33)	65 (±44)	55 (±32)
Heptachlor	52	86 (±44)	130 (±34)	110 (±31)	100 (±27)
Aldrin	20	84 (±38)	120 (±17)	110 (±23)	97 (±21)
Heptachlor epoxide	50	83 (±28)	120 (±17)	110 (±19)	100 (±17)
oxy-Chlordane	20	**	**	**	**
trans-Chlordane	30	83 (±45)	130 (±19)	110 (±15)	100 (±18)
o,p-DDE	67	83 (±29)	130 (±19)	110 (±17)	100 (±17)
Endosulfan I	210	82 (±26)	120 (±20)	110 (±18)	99 (±18)
cis-Chlordane	52	83 (±29)	120 (±20)	110 (±19)	100 (±17)
Dieldrin	45	83 (±23)	120 (±16)	110 (±15)	100 (±10)
p,p'-DDE	50	83 (±30)	130 (±19)	110 (±17)	100 (±20)
o,p-DDD	75	82 (±23)	120 (±15)	110 (±16)	100 (±15)
Endrin	63	87 (±30)	130 (±28)	110 (±24)	110 (±20)
Endosulfan II	79	80 (±23)	120 (±25)	100 (±16)	95 (±14)
p,p'-DDD	63	81 (±27)	120 (±18)	110 (±16)	100 (±16)
o,p-DDT	79	85 (±39)	120 (±28)	110 (±21)	100 (±27)
Endosulfan sulfate	120	76 (±17)	110 (±32)	100 (±21)	91 (±17)
p,p'-DDT	140	87 (±41)	130 (±33)	110 (±24)	110 (±34)
Methoxychlor	200	87 (±45)	130 (±42)	110 (±31)	110 (±42)

**   internal standard

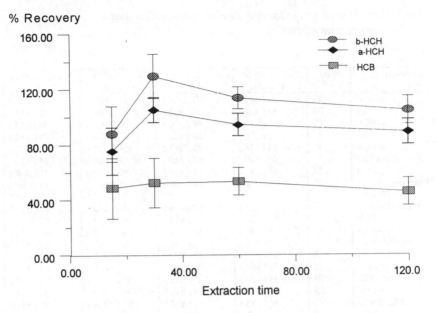

Figure 5. Variation of % recovery on extraction time by using general calculation

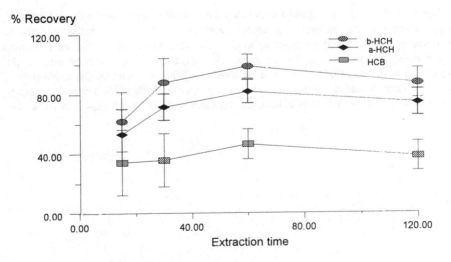

Figure 6

Table 3. Experimental design for optimization of extraction time
by using recovery factor.

Chemical	Spike (ng/L)	%R.(±%RSD) 15 mins	%R.(±%RSD) 30 mins	%R.(±%RSD) 60 mins	%R.(±%RSD) 120 mins
α-HCH	15	34 (±17)	72(±9.8)	82 (±7.8)	75(±9.2)
HCB	20	53 (±22)	36 (±13)	46 (±9.8)	38 (±10)
β-HCH	160	61 (±19)	88 (±16)	99 (±8.2)	88 (±10)
γ-HCH	32	52 (±20)	68 (±12)	82 (±7.8)	70 (±12)
δ-HCH	50	38 (±38)	43 (±17)	56 (±10)	46 (±12)
Heptachlor	52	60 (±30)	87 (±16)	93 (±10)	84 (±10)
Aldrin	20	59 (±27)	84 (±20)	95(±6.9)	82 (±8.2)
Heptachlor epoxide	50	58 (±35)	84 (±30)	97(±6.1)	85 (±5.4)
oxy-Chlordane	20	**	**	**	**
trans-Chlordane	30	59 (±28)	87 (±14)	96 (±12)	84 (±12)
o,p-DDE	67	59 (±20)	86 (±13)	96 (±6.7)	85 (±11)
Endosulfan I	210	58 (±24)	84 (±10)	96 (±6.0)	84 (±12)
cis-Chlordane	52	59 (±29)	85 (±16)	95 (±6.5)	84 (±11)
Dieldrin	45	59 (±25)	85 (±9.5)	98 (±6.8)	86(±9.0)
p,p'-DDE	50	59 (±27)	86 (±16)	95 (±7.2)	85(±11)
o,p-DDD	75	58 (±20)	85 (±18)	97 (±6.9)	85(±8.8)
Endrin	63	61 (±23)	87 (±12)	98 (±9.8)	91(±8.4)
Endosulfan II	79	56 (±23)	81 (±12)	92 (±6.5)	80 (±11)
p,p'-DDD	63	57 (±22)	83 (±11)	97 (±6.4)	85 (±13)
o,p-DDT	79	60 (±35)	85 (±15)	93 (± 10)	88 (±14)
Endosulfan sulfate	120	54 (±13)	74 (±13)	87 (±7.8)	76 (±10)
p,p'-DDT	140	61 (±33)	86 (±20)	95 (± 12)	97 (±9.8)
Methoxychlor	200	62 (±34)	86 (±20)	92 (± 14)	93 (±15)

The recovery factor can used to determine the concentration in sample in case of RSD <10 % such as δ- HCH and HCB that have %recovery in the range 40 -50 % and RSD < 10 %.

The recovery factor was applied to develop the analysis of organochlorine pesticides by using microextraction technique to compensate the volume in the extraction and to determine concentration of target compounds. The recovery factor was used to correct for determine the optimum of time consume and %recovery. The detection limit of this method are at the range of 0.02-0.1 ppb as shown in Table 4. However the application of recovery factor should depend on the chemical properties such as decomposition during extraction or injection that result in low recovery and % RSD should less than 10 %.

Table 4. Average percent recoveries (±%RSDs) and method detection
limit (MDL)of 22 organochlorine pesticide compounds
by using microextraction technique .

Chemical	Spike (ng/L)	%R.(±%RSD) (n=12)	MDL (ng/L)
α-HCH	6.1	79 (± 6.2)	1.4
HCB	8.2	40 (± 6.1)	1.0
β-HCH	63	73 (± 9.3)	21
α-HCH	13	73 (± 5.9)	2.7
δ-HCH	20	33 (± 5.6)	9.3
Heptachlor	21	97 (± 10)	3.1
Aldrin	13	84 (± 5.7)	9.3
Heptachlor epoxide	12	91 (± 5.4)	2.0
trans-Chlordane	12	80 (± 6.0)	2.8
o,p-DDE	27	85 (± 5.9)	6.7
Endosulfan I	84	92 (± 5.3)	7.6
cis-Chlordane	21	87 (± 5.9)	5.2
Dieldrin	18	89 (± 6.2)	4.9
p,p'-DDE	20	92 (± 6.9)	6.2
o,p-DDD	30	87 (± 5.4)	7.0
Endrin	25	96 (± 2.1)	2.0
Endosulfan II	32	91 (± 6.3)	23
p,p'-DDD	25	90 (± 5.5)	6.1
o,p-DDT	32	93 (± 0.9)	1.0
Endosulfan sulfate	49	84 (± 9.4)	18
p,p'-DDT	55	100 (± 2.5)	4.8
Methoxychlor	81	98 (± 4.8)	13

6. REFERENCES

1. Glaze W. H., R. RAWLEY., J. L. Burleson., D. Mapel and Dr. Scott,   *Adv. Identif. Anal. Org. Pollut. Water (Keith ; L.H.) , Inst. Appl. Sci. Texas , 1981, pp.267-280.*
2. G. H. Jeffery, J. Bassett, J. Mendham, R. C. Denney. "Vogel's Textbook Of Quantitative Chemical Analysis" Fifth edition, 1989.
3. Junk G.A., I. Ogawa and H. J. Svac "Extraction of Organic Compounds from Water using Small Amount of Solvent" Adv. Identif. Anal.Org. Pollut. Water (Keith ; L.H.) , Inst. Appl. Sci. Texas , 1981, pp.281-292.
4. Macek . K . J & Korn , S. (1970) J. FISH. Board Canada 27,1496-1498.
5. M. M. Thomson and W. Bertsch. *J. Chromatogr. 1983, 279, pp.383-393.*
6. Morrison, G. H. and H. Freisher, Solvent Extraction in Analytical Chemistry , John Wiley & Sons Inc., U.S.A.,1975.
7. Murry, D. A. J. *J. Chromatograph.,1979, 177, pp.135-140.*
8. Ramesh, A., Tanabe. S., Murase, H., Subramanian&Tatsukawa R. Environment . Poll . 74 (1991) 293-307.
9. Rhodes J. W. and C. P. Nulton , "Microextraction as an Approach to analysis for priority pollutions in Industrial Waste water" Adv. Identif. Anal. Org. Pollut. Water (Keith ; L.H.) ,pp.241 Inst. Appl. Sci.Texas , 1981, 241-252.
10. Thrun., K.E., K.E. Simmons and J.E. Oberhtzer.   *J. Environmental science and Health,*1980, 415, 5 pp. 485-501.
11. Thielen Dr., G. Olesen., A. Davis., E. Bajor., J. stefanovski and J.Chodkowski. *J. Chromatogr. Sci.,* 1987, 25, pp.12-16.

Analytical Recoveries and Their Use for Correction

J. F. Kay

VETERINARY MEDICINES DIRECTORATE, WOODHAM LANE, NEW HAW, ADDLESTONE, SURREY KT15 3NB, UK

1. INTRODUCTION

Analytical methods rarely give absolute accuracy and precision and the field of veterinary drug residue analysis is no exception. Methods commonly give rise to an answer lower than might be anticipated and with a degree of variability around this value. The ratio of this lower answer to the anticipated value expressed as a percentage is the overall recovery of the method and this is normally expressed as a mean value of a number of repeated analyses and the variability of these results either as a standard deviation or a relative standard deviation (also known as a coefficient of variation). Ideally recovery should be high with a low standard deviation but where this is not achievable it is normally regarded as more desirable to have a low recovery with a low deviation than a high recovery with a high deviation. Each step in the analytical sequence, e.g. extraction, will contribute to the overall recovery and a step wise consideration of the individual contributions can lead to improved overall recoveries.

Any analytical method may give rise to different recoveries and standard deviations depending on circumstances. For example a method could be applied to a series of identical samples in a single batch by a single analyst and give rise to a recovery and standard deviation. The same samples could be analysed in a number of batches by a single analyst and produce different recovery and precision values. Different analysts in the same or separate laboratories would be expected to achieve different recoveries and precisions. A further set of recoveries and precision values is possible under all of the above circumstances if the analyses are performed on samples containing different concentrations of the analyte.

Validation of a method of analysis can be performed under all of these circumstances viz:

a. a batch of identical samples by a single analyst on one occasion;
b. a batch of identical samples by a single analyst on a number of occasions;
c. batches of identical samples by different analysts within a laboratory;
d. batches of identical samples by different analysts in different laboratories.

To date results of reported veterinary drug analyses fall predominantly into categories a, b and c. Few methods have undergone the rigorous collaborative trialling required by category d.

2. DETERMINATION OF ACCURACY AND PRECISION

The accuracy and precision of a method are normally determined by spiking a known quantity of the analyte into a blank control sample. This is most commonly done at the start of the analysis although later addition is sometimes used. Final detector response to this quantity of analyte is determined and the response obtained from the 'spiked' sample is expressed as a percentage of this to give a recovery. Recoveries from multiple spiked samples then give rise to an average recovery and the deviation about this. This method is known as using an external standard.

An alternative approach is to use an internal standard ideally using a substance as closely related to the analyte in question as possible but nevertheless separately determinable and unlikely to occur in any sample. For example thiamphenicol has been used as an internal standard for chloramphenicol[1] and democlocyline has been used in tetracycline[2] analysis.

Where such an approach is used, it is normal to determine in 'spiked' samples the recoveries of both the analyte and the internal standard repeatedly to obtain mean recoveries of both thus allowing a ratio to be established for the analytical procedure, e.g. if thiamphenicol (T) has a mean recovery of 75% and chloramphenicol (C) 60% the ratio of T/C is 75/60 or 1.25. By adding a standard quantity of internal standard to every surveillance sample, where an analyte response is obtained the internal standard response allows the analyte response to be calculated to the equivalent of 100% recovery.

The ideal internal standard is a labelled form of the analyte itself such as a multiply deuterated molecule. Chemically such a molecule should behave identically to its non-deuterated form and only vary in minor physical characteristics. So long as the two forms can be separately determined at the end step, the labelled molecule makes an ideal internal standard.   Means of determination include chromatographic separation   relying on molecular weight differences; measurement of different masses by mass spectrometry; and, measuring radioactivity if the labelled form of the molecule contains such an isotope e.g. tritium. Examples of where such an approach is used are D6clenbuterol in the [1]-agonists determination[3] or radio immunoassay of hormones when tritiated forms of the anabolics[4] are used. This approach again gives rise to 100% recoveries because regardless of the actual recovery of the labelled compounds (since this is added in known quantity) the response of the analyte to be determined can be expressed as the determined ratio of this quantity.

3. USE OF THE RECOVERY FIGURE

During surveillance a spiked blank sample is normally included with each batch of samples to allow the recovery of the method to be determined. The analyst is left with the choice of quoting all results "as is" with the recovery value given or correcting the "as is" figure by use of the recovery data obtained.

Recovery is commonly less than 100% so correction of an "as is" result will inflate this result and where recovery of a method is poor (* 50%) this could result in a doubling or more of the "as is" value.

Normally recovery experiments are carried out in parallel with surveillance samples. However it must be remembered that recovery values are only obtained for "spiked or added" analyte and not on normally incurred tissue unless animals have been treated with radiolabelled drugs prior to matrix collection. The incurred tissue may not give a similar recovery because the analyte may be "bound" into the matrix either ionically or covalently. Correction using the spiked sample recovery value will still result in a low value.

Since Maximum Residue Limits (MRLs) are now in force for a significant number of analyte/matrix combinations under European Union regulations,[5] a resolution to the question of whether or not to correct for recovery is particularly important. If correction for analytical recovery is the way forward, as required by European Commission Decision 93/256/EEC[6] for residues of substances having a hormonal or thyrostatic action, it is imperative to agree on the way in which such factors should be derived and applied. An internationally agreed approach to this issue is obviously the best way forward. Nationally, correction of an "as is" result below the MRL could bring it above this value resulting in condemnation of the carcass and court action. Internationally, the implications of the previous case could result in sanctions against trading partners.

There is considerable logic in correcting all results for recovery. In the EU, MRLs are set following the establishment of an Acceptable Daily Intake (ADI) which is derived from actual doses of the substance of interest. In setting the MRL, attention is paid to a suitable marker residue which may be the parent substance, a metabolite, or a combination of both. Therefore, any residue determined should represent the total residue of concern available and not merely the raw data determined by the analyst.

3.1 Consequences of Correction

The consequences of correction are that:-
    a. The consumer receives additional protection where action is permissible;
    b. The consumer becomes more concerned since more samples will be found positive and others will be reported with higher values;
    c. The reader is not required to do anything.

3.2 Consequences of Non-Correction.

The consequences of non-correction are the opposite of the above, i.e.:-

    a. The producer receives additional protection
    b. Fewer samples are found positive and those which are will generally be at lower values;
    c. The reader has the option to use the recovery figure where quoted to convert the "as is" figure into a corrected value.

4. REPORTING OF RESULTS BY THE VMD

The VMD controls two programmes of veterinary drug residue surveillance, one statutory and the other non-statutory. As an indication of the scale of these programmes, approximately 63,000 analyses were scheduled to be conducted in 1995 and the results regularly updated in MAVIS.[7] To date, the policy of the VMD is to quote all analytical results for statutory surveillance as found and provide a recovery figure which the reader may use if he/she so wishes to produce a corrected value. However, the data for the non-statutory surveillance is reported as corrected for analytical recovery.

It has become clear that divergent approaches have been adopted in more recent times internationally, with some data being presented corrected for recovery while other data remained uncorrected. The discussions at this meeting will inform the debate within the VMD on the need for consistency of approach.

5. REFERENCES

1. H. J. Keukens, W. M. J. Beek and M. M. L. Aerts, J. Chromatogr., 1986, 352, 445.
2. W. H. H. Farrington, J. Tarbin, J. Bygrave and G. Shearer, Food Add. Contam., 1991, 8(1), 55-64.
3. J. M. Sauer, R. J. H. Pickett, S. N. Dixon and R. Jackman, Proceedings of EuroResidue III, 3-5 May 1993, 622-626.
4. W. J. McCaughey, Veterinary Record, 1988, 123, 511-513.
5. European Community Council Regulation No. 2377/90. Official Journal of the European Communities, 1990, L224, 1.
6. European Community Commission Decision 93□ü¿/256/EEC. Official Journal of the European Communities, 1993, L118, 64.
7. Medicines Act Veterinary Information Service (MAVIS), issued quarterly by the VMD.

codex alimentarius commission

FOOD AND AGRICULTURE
ORGANIZATION
OF THE UNITED NATIONS

WORLD HEALTH
ORGANIZATION

JOINT OFFICE: Via delle Terme di Caracalla 00100 ROME   Tel.: 52251   Telex: 625825-625853 FAO I   Cables: Foodagri Rome   Facsimile: (6)5225.4593

ALINORM 97/23

JOINT FAO/WHO FOOD STANDARDS PROGRAMME

CODEX ALIMENTARIUS COMMISSION
Twenty-second Session
Geneva, 23-28 June 1997

REPORT OF THE 20TH SESSION OF THE
CODEX COMMITTEE ON METHODS OF ANALYSIS AND SAMPLING
Budapest, Hungary, 2-6 October 1995

Note: This report incorporates Codex Circular Letter CL 1995/46-MAS.

codex alimentarius commission

FOOD AND AGRICULTURE   WORLD HEALTH
ORGANIZATION   ORGANIZATION
OF THE UNITED NATIONS

JOINT OFFICE: Via delle Terme di Caracalla 00100 ROME   Tel.: 52251   Telex: 625825-625853 FAO I   Cables: Foodagri Rome   Facsimile: (6)5225 4593

CX 4/50.2

CL 1995/46-MAS
October 1995

To: - Codex Contact Points
             - Participants at the Twentieth Session of the
               Codex Committee on Methods of Analysis and Sampling
             - Interested International Organizations

From: Chief, Joint FAO/WHO Food Standards Programme, FAO
             Vialle delle Terme di Caracalla, 00100 Rome, Italy

Subject: Distribution of the Report of the Twentieth Session, of the Codex Committee on
             Methods of Analysis and Sampling (CCMAS).

       The report of the Twentieth Session of the above Committee (ALINORM 97/23) will be
considered by the Twenty-second Session of the Codex Alimentarius Commission (Geneva, 23-28
June 1997).

PART A: MATTERS FOR ADOPTION BY THE COMMISSION

       The following matters will be brought to the attention of the 22nd Session of the Codex
Alimentarius Commission for adoption:

i.   The IUPAC/ISO/AOAC Harmonized Guidelines for Internal Quality Control in Analytical
       Chemistry Laboratories (ALINORM 97/23, para. 40) and

ii.   Amendment to the Committee's Terms of Reference (ALINORM 97/23, para. 64).

       The Committee also endorsed provisions concerning methods of analysis for 10 commodity
Codex Standards and also assigned type classification to 12 Codex General Methods for Contaminants
(ALINORM 97/23, para 54 and Appendix IV).

       Governments wishing to propose amendments or to submit comments regarding the implications
which the above matters have for their economic interest should do so in writing, in conformity with
the Codex Alimentarius Commission Procedural Manual, to the Chief, Joint FAO/WHO Food
Standards Programme, FAO, Vialle delle Terme di Caracalla, 00100 Rome, Italy, **no later than 31st**
April 1996.

SUMMARY AND CONCLUSIONS

The Twentieth Session of the Codex Committee on Methods of Analysis and Sampling reached the following conclusions:

MATTERS FOR CONSIDERATION BY THE COMMISSION AND THE EXECUTIVE COMMITTEE:

- Recommended the adoption of the IUPAC/ISO/AOAC Harmonized guidelines for internal quality control in analytical chemistry laboratories (paras. 37-40);

- Recommended amendments to the Committee's Terms of Reference (paras. 63 & 67);

OTHER MATTERS OF INTEREST TO THE COMMISSION

- Agreed in principle to accept the criteria approach for evaluating methods of analysis for Codex purposes (para. 17);

- Agreed that the following quality criteria be adopted by laboratories involved in the official import and export control of foods:
  - Comply with the general criteria for testing laboratories laid down in ISO/IEC Guide 25:1990
  - Participate in appropriate proficiency testing schemes for food analysis which conform to the requirements laid down in "The International Harmonized Protocol for the Proficiency testing of (Chemical) Analytical Laboratories",
  - Whenever available, use methods of analysis which have been validated according to the principles laid down by the Codex Alimentarius Commission; and
  - Use internal quality control procedures, such as those described in the "Harmonized Guidelines for Internal Quality Control in Analytical Chemistry Laboratories" (para. 21);

- Recommended the adoption of the IUPAC/ISO/AOAC Harmonized Protocol for the use of Recovery Factors for Codex purposes when the protocol was published by IUPAC (para. 27);

- Agreed to request Commodity Committees to identify how extensive the problem of indirect determinations was in Codex Standards (para. 30);

- Agreed to define selected core terms of direct relevance to the work of Codex and circulate to governments and international organizations for comments (para. 34);

- Noted the report of the Inter-Agency Meeting (para. 47);

- Noted the progress report on review of standard methods of analysis and sampling (para. 48);

- Requested commodity Committees to avoid selecting methods of analysis which use ozone-depleting solvents (paras. 61 & 62) and

- Proposed to undertake the following new work:

  - Review of methods of analysis using ozone-depleting substances and
  - Measurement uncertainty (para. 66).

TABLE OF CONTENTS

Paragraphs

INTRODUCTION   1

OPENING OF THE SESSION   2

ADOPTION OF THE AGENDA   3

APPOINTMENT OF RAPPORTEUR   5

PROPOSED DRAFT CODEX GENERAL GUIDELINES ON SAMPLING   6 - 11

CRITERIA FOR EVALUATING ACCEPTABLE METHODS OF ANALYSIS
  FOR CODEX PURPOSES   12 - 18

DEVELOPMENT OF OBJECTIVE CRITERIA FOR ASSESSING THE
  COMPETENCE OF TESTING LABORATORIES INVOLVED IN THE
  OFFICIAL IMPORT AND EXPORT CONTROL OF FOODS   19 - 23

PROGRESS REPORT ON THE DEVELOPMENT OF AN
  IUPAC/ISO/AOAC HARMONIZED PROTOCOL FOR RECOVERY FACTORS   24 - 27

DEVELOPMENT OF UNIFORM CRITERIA FOR THE REPORTING OF
  TEST RESULTS ESPECIALLY WHEN THE PROVISION OR SPECIFICATION
  TO BE TESTED IS NOT IDENTICAL TO THE ANALYTE   28 - 30

HARMONIZATION OF ANALYTICAL TERMINOLOGY IN ACCORDANCE WITH
  INTERNATIONAL STANDARDS   31 - 36

PROGRESS REPORT ON THE DEVELOPMENT OF THE IUPAC/ISO/AOAC
  HARMONIZED PROTOCOL FOR THE QUALITY CONTROL OF (CHEMICAL)
  ANALYTICAL DATA   37 - 40

REPORT OF THE ELEVENTH MEETING OF INTERNATIONAL ORGANIZATIONS
  WORKING IN THE FIELD OF METHODS OF ANALYSIS AND SAMPLING
  (INTER-AGENCY MEETING), AND PROGRESS REPORT ON REVIEW OF
  STANDARD METHODS BY INTERNATIONAL ORGANIZATIONS   41 - 50

ENDORSEMENT OF METHODS OF ANALYSIS IN CODEX STANDARDS   51 - 55

REVIEW OF METHODS OF ANALYSIS USING OZONE-DEPLETING SUBSTANCES   56 - 62

OTHER BUSINESS AND FUTURE WORK   63 - 67

DATE AND PLACE OF THE NEXT SESSION   68

Page

SUMMARY STATUS OF WORK   14

TABLE OF CONTENTS (Cont'd)

APPENDICES

		Page
Appendix I	List of Participants	15
Appendix II	Harmonized Guidelines for Internal Quality Control in Analytical Chemistry Laboratories	25
Appendix III	Amendment of the Terms of Reference of the Codex Committee on Methods of Analysis and Sampling	26
Appendix IV	List of Methods of Analysis considered by the Twentieth Session of the Codex Committee on Methods of Analysis and Sampling	27

ALINORM 97/23

REPORT OF THE TWENTIETH SESSION OF THE
CODEX COMMITTEE ON METHODS OF ANALYSIS AND SAMPLING
Budapest, Hungary, 2-6 October 1995

INTRODUCTION

1.   The Codex Committee on Methods of Analysis and Sampling held its Twentieth Session from 2 to 6 October 1995 in Budapest, by courtesy of the Government of Hungary. The Session was chaired by Professor Peter Biacs, Director General of the Central Food Research Institute (KEKI). The Session was attended by 109 delegates and observers from 41 countries and 5 International Organizations. A complete list of participants, including the Secretariat is provided in Appendix I to this report.

OPENING OF THE SESSION (Agenda Item 1)

2.   At the opening session, the delegates were welcomed by Dr. E. Rácz, Director of Food Quality Division, Department of Food Industries and current Chairman of the Hungarian Codex Committee. The session was addressed by Dr. Lászlo Vajda, Head of the Department of International Economic Relations, Ministry of Agriculture.

ADOPTION OF THE AGENDA[1] (Agenda Item 2)

3.   The Committee adopted the Provisional Agenda as proposed, and agreed to:

-       appoint an *ad hoc* Working Group to consider Agenda item 4 - Proposed Draft Codex General Guidelines on Sampling, in order to facilitate its consideration;

-       discuss Agenda item 7 directly after Agenda item 5;

-       discuss, "Review of Methods of Analysis Using Ozone-Depleting Substances" after Agenda item 12; and

-       discuss amendment to the Terms of Reference of the Committee under Agenda item 13 - Other Business and Future Work.

4.   Some delegations requested addition of an Agenda item on "Matters of interest".   The Chairman informed the Committee that the Commission at its 21st Session approved the Committee's proposal for new work[2] and adopted the three items (two Protocols and the Codex Methods of Analysis for Contaminants).

APPOINTMENT OF RAPPORTEUR (Agenda Item 3)

5.   The Committee agreed with the proposal of the Chairman to appoint Mr. William J. Franks, (USA) as rapporteur.

---

[1]   CX/MAS 95/1
[2]   ALINORM 95/4, para. 8 & Appendix II; ALINORM 95/37, para. 12

PROPOSED DRAFT CODEX GENERAL GUIDELINES ON SAMPLING[3] (Agenda Item 4)

6.   It was noted that since the last Session of the Committee, the proposed draft guidelines had been circulated for comments and, taking into consideration the comments received, revised by a Consultant, Dr. R. Coker[4]. Dr. Coker presented the revised paper and emphasized that the main aim of the revision was to make the document more user-friendly and comprehensive.

7.   After the introduction of the document, an *ad hoc* Working Group was convened in order to facilitate the discussion of the document. It was chaired by Dr. F. McClure of the USA and was composed of delegates from Canada, Finland, France, Hungary, The Netherlands, Norway and the USA. Dr. Coker was the rapporteur.

8.   After much deliberation, the Working Group recommended to the plenary that the document should be further revised. The revised document would consist of two main parts as follows:

Part I: DISCRETE LOTS MOVING IN INTERNATIONAL TRADE

(a)   Two class attributes plans for proportion of non-conforming units;

(b)   Three class attributes plans (for microbiological assessments);

(c)   Variables plans for proportion of non-conforming units: <u>unknown</u> standard deviation; and

(d)   Attributes plans to detect at least one non-conforming unit in a lot.

Part II: CONTROL OF MANUFACTURING PROCESS

(a)   Two class attributes plans for proportion non-conforming units (ISO 2859);

(b)   Variables plans for proportion of non-conforming units: <u>known</u> and <u>unknown</u> standard deviation (ISO 3951);and

(c)   Switching Rules.

9.   Additional changes would include:

•   Use ISO definitions throughout the document (ISO 7002 as primary source);

•   Include additional AQLs (0.16, 0.4, 1.6); and

•   Make a number of other changes, such as removing Section 5.1.1, "The treatment of lots of varying size" and the associated Table 7.

10.   The Committee agreed that the Secretariat should arrange for the present draft guidelines to be further revised, with review assistance provided by the members who served on the *ad hoc* Working Group. A new revised draft should be circulated to Member Countries for comments at

---

3   CX/MAS 95/2
       Comment papers (Czech Republic & Hungary)
4   Ray Coker, Ph.D., Principal Natural Products Scientist, Natural Resources Institute, Chatham, UK.

Step 3 well before the next Session of the Committee. This revised draft should identify potential users of the document.

11.   It was also suggested that the current document with the suggested revision should be brought to the attention of the Codex Committees on Residues of Veterinary Drugs in Foods and on Food Hygiene during their Sessions in November/December 1995, with an indication that the document is under revision, and that the revised document should be presented to the Codex Committee on Pesticide Residues which will meet next year.

CRITERIA FOR EVALUATING ACCEPTABLE METHODS OF ANALYSIS FOR CODEX PURPOSES[5] (Agenda Item 5)

12.   The Committee recalled that this item had been discussed at the two previous sessions without reaching an agreement. The Delegation of the United Kingdom presented the paper. It was stated that in order to overcome disadvantages of the current system and to give analysts freedom of choice, an alternative approach was proposed - to define criteria and to choose methods which met criteria instead of specifying individual methods. The Committee noted that in the new approach Types I and IV would remain as at present whereas Types II and III would be converted into criteria.

13.   The majority of delegations were in favour of this new approach. Nonetheless, several delegations foresaw the enormity of the task to convert these methods into criteria, including selection of criteria. Some delegations preferred to include other criteria such as "accuracy" ("trueness" or "bias"). It was pointed out that some criteria for selecting methods had already been adopted by the Commission[6].

14.   It was stressed that the methods which met criteria should be collaboratively studied and validated according to the protocol on inter-laboratory studies. Some delegations noted that an external standard, such as Horwitz curve, should be applied. The Delegation of Hungary stated that sample preparation be taken into consideration in addition to measurement of analytes.

15.   The Delegation of the USA pointed out some discrepancies in Appendices I and II. The Delegation of the UK responded that these problems represent limitations of the current system of selecting methods.

16.   The Delegation of the USA stated while it could accept this new approach for Type III methods, it was strongly opposed to the application of this approach to Type II methods. In the case of disputes, only one method should be chosen (current Type II methods) and used by all parties involved, especially when a dispute becomes a legal or administrative issue. It was stressed that the Committee should retain the prerogative to determine Type II methods.

17.   The Committee agreed to recommendation 1[7], to accept the approach in principle. The Committee also agreed to proceed along the line set out in the other recommendations with the understanding that there should be a clear indication that the problems related to TypeII/TypeIII

---

5   CX/MAS 95/3

6   Comment papers (USA, IDF & IUPAC
       "Recommendations for a Checklist of Information Required to Evaluate Methods of Analysis Submitted to
       the Codex Committee on Methods of Analysis and Sampling for Endorsement", *Codex Alimentarius*, Second
       Edition, Volume 13, pp. 129, and "Methods of Analysis Submitted for Endorsement by the Codex
       Committee on Methods of Analysis and Sampling: Precision Criteria", ALINORM 93/23, Appendix III.

7   Page 5, CX/MAS 95/3

classification were deliberately not dealt with. The Committee agreed to separate recommendation 3 into two i.e. new 3 and 4. The recommendations are cited below:

1. Accept the criteria approach in principle;

2. Draw up detailed working guidelines for the operation of the criteria approach by CCMAS. This would include the definitions and selection of the criteria to be used;

3. Clarify the procedures to be used in the 'dispute situation'; and

4. Emphasise that procedures are to be used to ensure that laboratories are 'in control' and operating proficiently in all cases.

18.   The Committee requested the Delegations of the United Kingdom and Canada in collaboration with the Codex Secretariat to prepare a paper on working procedures for the new approach in horizontal manner, using Codex General Methods for Contaminants as examples, for consideration by the Committee at its next session and by its *ad hoc* Working Group on Endorsement. The Committee invited other delegations to make contribution.

DEVELOPMENT OF OBJECTIVE CRITERIA FOR ASSESSING THE COMPETENCE OF TESTING LABORATORIES INVOLVED IN THE OFFICIAL IMPORT AND EXPORT CONTROL OF FOODS[8] (Agenda Item 6)

19.   The document was prepared and introduced by the Delegation of Finland. It was emphasized that ISO/IEC Guide 25:1990 should form the basis of objective criteria for assessing the competence of testing laboratories involved in the import and export control of foods. In addition, such laboratories should participate in proficiency testing, and use validated methods.   Some delegations preferred the deletion of the word "official" from the title of the text, while others preferred that testing laboratories remained within the framework of official control.   The Committee agreed to include the word "official" in the title of the text.   Some delegations stated that the reference to ISO/IEC Guide 25:1990 was sufficient and there was no need to refer to proficiency testing. However, it was pointed out that ISO/IEC Guide 25:1990 did not specifically address participation in proficiency testing in the field of food analysis, which is required to demonstrate competence in this field.

20.   The Committee agreed to provide more general wording instead of referring to "third party" in the text and to include mailing addresses for each reference cited.

21.   The Committee agreed that the following quality criteria be adopted by laboratories involved in the official import and export control of foods:

-   Comply with the general criteria for testing laboratories laid down in ISO/IEC Guide 25:1990 "General requirements for the competence of calibration and testing laboratories"[9],

-   Participate in appropriate proficiency testing schemes for food analysis which conform to the requirements laid down in "The International Harmonized Protocol for the Proficiency Testing of (Chemical) Analytical Laboratories", *Pure & Appl. Chem.* **65** (1993) 2132-2144

---

[8]   CX/MAS 95/4.
       Comment papers (USA, IDF & IUPAC)
[9]   Currently under revision.

(as adopted for Codex purposes by the Codex Alimentarius Commission at its 21st Session in July 1995);

- Whenever available, use methods of analysis which have been validated according to the principles laid down by the Codex Alimentarius Commission; and

- Use internal quality control procedures, such as those described in the "Harmonized Guidelines for Internal Quality Control in Analytical Chemistry Laboratories", *Pure & Appl. Chem.* 67 (1995) 649-666[10].

22.   The Committee noted that compliance with the criteria mentioned for laboratories involved in the official import and export control of foods needed to be assessed by suitable mechanisms. The bodies assessing the laboratories should comply with the general criteria for laboratory accreditation, such as those laid down in ISO/IEC Guide 58:1993, "Calibration and testing laboratory accreditation systems - General requirements for operation and recognition".

23.   It was **agreed** that the paper be revised, based on the comments and recommendations made during the session. Noting the work currently carried out by the Codex Committee on Food Import and Export Certification and Inspection Systems in the area of import and export control in general, the Committee also agreed that the revised paper should be referred to the Codex Committee on Food Import and Export Certification and Inspection Systems for its consideration, review and comments.

PROGRESS REPORT ON THE DEVELOPMENT OF AN IUPAC/ISO/AOAC HARMONIZED PROTOCOL FOR RECOVERY FACTORS[11] (Agenda Item 7)

24.   The paper was prepared and presented by the Delegation of the United Kingdom. The Committee was informed that the paper was comprised of a collection of informal discussions by analysts on the issue of recovery factors. The application of recovery factors was of particular interest especially where the difference between a corrected and uncorrected result affects a product's compliance with a specification provision.

25.   The Committee noted that the Inter-Divisional Working Party of IUPAC was preparing questionnaires requesting information on the status of applying recovery factors. The Delegation of the United Kingdom requested that other delegations provide it comments concerning this initiative of IUPAC. It was noted that results of the survey would be the basis for an ISO/IUPAC/AOAC symposium, organized by AOAC International to be held during its annual meeting in Orlando, Florida. The Committee was informed that the protocol on recovery factors to be developed from the conclusions of the symposium might be published in 1998.

26.   Many delegations felt the use of recovery factors to be an important topic. The Committee was informed that the paper did not address the factor of propagation of errors when using recovery factors. It was pointed out that some methods, such as those for residues of veterinary drugs and pesticides did not require correction for recovery. Recoveries had already been considered in setting up the maximum residue limit for the veterinary drug or pesticide as appropriate. It was also pointed out that it was necessary to know the nature of the analyte, whether it was free or non-free.

---

10   Being recommended to the Commission for adoption (see para. 40).
11   CX/MAS 95/5
        Comment papers (IDF & IUPAC)

27.   The Committee indicated its interest in the work on the use of recovery factors being undertaken by IUPAC. The Committee requested to be kept informed by IUPAC of the progress being made on the development of an IUPAC/ISO/AOAC Harmonized Protocol for the Use of Recovery Factors. In the future, the Committee might recommend the document to the Commission for adoption by reference for Codex purposes when the protocol was published by IUPAC.

DEVELOPMENT OF UNIFORM CRITERIA FOR THE REPORTING OF TEST RESULTS ESPECIALLY WHEN THE PROVISION OR SPECIFICATION TO BE TESTED IS NOT IDENTICAL TO THE ANALYTE[12] (Agenda Item 8)

28.   The Committee noted that the Commission at its 21st Session approved the initiation of this work that had been proposed by the Delegation of Austria at the last Session of the Committee. The Delegation of Austria submitted its comments and draft guidelines for reporting analytical results at the Session which contained sections on: name of the parameter; additional information; value and unit; and limit of detection or limit of determination/quantification (in relation to negative results). The Delegation proposed to elaborate the guidelines. The Committee noted that the IUPAC was currently investigating how to report low level results including negative signals and matters related to values and units.

29.   It was felt that there were no significant problems for this Committee and therefore, no need to draw up guidelines. If there were problems, Commodity Committees were in better position than this Committee to identify them and they could solve the problems by modifying the specifications in the standards or request guidance from this Committee. When a provision or specification to be tested was not identical to the analyte, how to express the analytical results should be clearly stated in the standard in order to avoid the problem.

30.   The Committee agreed to request Commodity Committees to identify how extensive the problem of indirect determinations was in Codex Standards. Based on responses from the Commodity Committees, the Codex Committee on Methods of Analysis and Sampling might consider elaborating guidance to these Commodity Committees, such as guidelines and appropriate factors. If there were no problems identified, the Committee should seek approval of the Commission to discontinue work in this area.

HARMONIZATION OF ANALYTICAL TERMINOLOGY IN ACCORDANCE WITH INTERNATIONAL STANDARDS (Agenda item 9)

31.   The Committee considered the paper prepared by AOAC INTERNATIONAL at the request of the Codex Secretariat. The paper provided a bibliography to assist the Committee in the harmonization of analytical terminology[13], and also made recommendations on how to progress.

32.   While the Committee recognized the usefulness of such a harmonized document, it was also noted that other bodies had abandoned the idea because of the enormity of the task.

33.   The Committee therefore agreed to limit the scope of the work involved to defining a small number of terms specifically related to the work of the Committee. The Committee considered that those terms as contained in the Codex Alimentarius Procedural Manual, and the harmonized protocols adopted by the Commission should be ones to be defined. The Committee accepted the

---

[12]   CX/MAS 95/6
          Comment paper (Austria)
[13]   CX/MAS 95/7

recommendations made in the paper. The Delegations of the United States and Finland accepted the Committee's request to undertake the assignment and AOAC, ISO and IUPAC were requested to collaborate with them.

34.   The Committee agreed to the proposal that the Delegations of USA and Finland in collaboration with AOAC INTERNATIONAL, ISO and IUPAC would, during the Session, provide core terms of direct relevance to the work of Codex, for inclusion in the report. Before the next session, the terms would be defined and circulated to governments and interested International Organizations. Based on the comments received, a revised paper would be prepared for consideration at its 21st Session and the IAM at its 12th meeting.

35.   The Committee considered the list prepared, as requested above, and provided comments which were utilized in producing the revised tentative list below:

TERMS PERTINENT TO THE CODEX COMMITTEE ON METHODS OF ANALYSIS AND SAMPLING WHICH NEED TO BE DEFINED

-        Terms included in the Procedural Manual of the Codex Alimentarius Commission
-        Terms included in the harmonized protocols
    -        for method-performance studies,
    -        for proficiency testing schemes, and
    -        for internal quality control

The Term Final Value

Uncertainty (Reliability) Terms
-        Accuracy (recovery)
    -        error of single value
    -        trueness
    -        bias
-        Precision (extremes)
    -        repeatability (intermediate)
    -        reproducibility

Method Characteristics Terms
-        applicability?
-        specificity
-        sensitivity
-        ruggedness (validation)
-        limits
    -        provision
    -        decision
    -        quantitation

Interlaboratory Studies Terms
-        Method-performance Studies (validation of methods, outliers, invalid data)
-        Laboratory-performance (Proficiency) Studies
-        Material-performance Studies (including various kinds of reference materials)

Terms Related to Method Types (I - IV)

36.   The Chairman thanked the Representative of AOAC INTERNATIONAL for the paper and also recognized the initiative of the Delegations of the United States and Finland.

PROGRESS REPORT ON THE DEVELOPMENT OF THE IUPAC/ISO/AOAC HARMONIZED PROTOCOL FOR THE QUALITY CONTROL OF (CHEMICAL) ANALYTICAL DATA[14] (Agenda Item 10)

37.   The Committee noted that at its 19th Session it had had for discussion the IUPAC/ISO/AOAC Harmonized Guidelines for Internal Quality Control in Analytical Chemistry Laboratories. As it had been informed that the harmonized guidelines had been planned to be further considered by IUPAC in May 1994, the Committee had agreed that after a revised document was adopted by IUPAC, the Committee should consider the document with a view towards adopting it for Codex purposes.

38.   The Committee was informed that IUPAC had adopted and published the finalized harmonized guidelines. Currently ISO and AOAC were reviewing the document with the goal of harmonization with IUPAC. The Delegation of the United Kingdom stated that the harmonized guidelines was the third of such documents that were elaborated by the IUPAC's Inter-divisional Working Party, the first two having been already adopted by the Codex Alimentarius Commission. The Delegation stressed that the document was intended to be of an advisory nature as opposed to a mandatory nature of protocol.

39.   The Delegation of Sweden suggested that it was sufficient to include the harmonized guidelines into the recommendations made to the Codex Committee on Food Import and Export Certification and Inspection Systems on development of objective criteria for assessing the competence of testing laboratories involved in the import and export control of foods (see para. 23). However, the majority of delegations preferred formal adoption by the Commission so that there would be guidance on internal quality control procedures available to Codex.

Status of the Harmonized Guidelines for Internal Quality Control in Analytical Chemistry Laboratories

40.   Recognising the advisory nature of the document, the Committee recommended the Harmonized Guidelines[15], to the Commission for adoption for Codex purposes.

REPORT OF THE ELEVENTH MEETING OF INTERNATIONAL ORGANIZATIONS WORKING IN THE FIELD OF METHODS OF ANALYSIS AND SAMPLING (INTER-AGENCY MEETING), AND PROGRESS REPORT ON REVIEW OF STANDARD METHODS BY INTERNATIONAL ORGANIZATIONS (Agenda Item 11)

(i) INTER-AGENCY MEETING

41.   The Report was presented by Mr. K.-G. Lingner (ISO), Secretary of the Inter-Agency Meeting (IAM). The IAM was attended by representatives of 11 international organizations (AOAC, CAC, CEN, EOQ, ICC, ICUMSA, IDF, ISO, IUPAC, NMKL and OIV) and was chaired by Mr. G. Castan (ISO).

---

[14]   CX/MAS 95/8
        Comment paper (IDF)
[15]   Appendix II of this Report

the Commission that the CCMAS should maintain the closest possible relationship with all interested organizations working on methods of analysis and sampling.[16]

46.   The Chairman said that the Host Government would be prepared to reproduce and circulate the Report of the IAM. Reflecting the views of the Committee, the Chairman said that he considered the IAM an integral part of the Committee and he would suggest that the Executive Committee discuss the importance of IAM at its next session.

47.   The Committee noted the report of the Inter-Agency Meeting and expressed its appreciation for the assistance the IAM was providing to it.

(II) PROGRESS REPORT ON REVIEW OF STANDARD METHODS BY INTERNATIONAL ORGANIZATIONS ON METHODS OF ANALYSIS AND SAMPLING

48.   The Committee also noted the progress[17] report by the Delegation of the United Kingdom on review of standard methods by international organizations on methods of analysis and sampling. Some delegations made useful suggestions to improve on the information provided in the report.

49.   The Committee was informed by the Delegation of the United Kingdom that the updated report would be available as an information paper at the next meeting of the IAM and the Session of the Committee.

50.   The Committee expressed its appreciation and requested the Delegation of the United Kingdom to continue the preparation of the report.

ENDORSEMENT OF METHODS OF ANALYSIS IN CODEX STANDARDS[18]
(Agenda Item 12)

51.   A report of the *ad hoc* Working Group on Endorsement was introduced by its chairman, Dr. W. Horwitz (USA).   Dr. G. Diachenko (USA) served as rapporteur.   The following Member Countries and International Organizations had been represented: Canada, Finland, France, Hungary, The Netherlands, Slovakia, Thailand, the United Kingdom, the United States, AOAC, IDF, ISO, IUPAC and OIV. The Group had considered: (i) Type of Codex General Methods for Contaminants adopted by the Commission at its 21st Session; (ii) Codex Methods of Analysis and Sampling (CAC/RMs); and (iii) Methods of Analysis for Commodity Standards (except those for sugars, fats and oils[19]).

52.   Concerning Codex Methods of Analysis and Sampling (CAC/RMs), it was recommended that the Commodity Committees be advised to consider replacing some of the methods with more modern ones as appropriate and replace the CAC/RM numbers with the original literature references, if possible. The Committee agreed to recommend to the Commission the deletion of the CAC/RM numbering system.   International organizations whose methods were contained in the list of CAC/RMs were invited to review their methods and to communicate any proposed updated reference citations to the Codex Secretariat as AOAC and ICUMSA had done.

---

16   Codex Alimentarius Procedural Manual, Eighth Edition, Page 108
17   Conference Room Document 3
18   CX/MAS 95/9, CX/MAS 95/9-Add. 1, and Conference Room Document 1.
19   See Appendix IV Notes.

42.   The IAM had considered matters of interest to Codex Committee on Methods of Analysis and Sampling, such as:

- international collaboration in the field of standard methods of analysis and sampling;
- methods of analysis and sampling required by the Codex Alimentarius Commission;
- proprietary laboratory techniques;
- ownership rights for methods and copyrights; and
- publication *in extenso* of a compendium of methods approved by the Codex Alimentarius Commission.

43.   As a result of its discussions, the IAM approved the following recommendations:

- that a summary of the IAM proceedings be included in the body of the Report of the Twentieth Session of Codex Committee on Methods of Analysis and Sampling;
- that the subject of quality assurance in food analysis be included in the agenda of the next IAM;
- that the document concerning proprietary laboratory techniques prepared by AOAC INTERNATIONAL be re-circulated and that all organizations participating in the IAM be invited to submit comments and information on their respective policies and practices to the IAM Secretariat;
- that the IAM Secretariat be requested to re-circulate a survey of member agencies of their procedures and practices concerning ownership rights (copyrights), including bilateral and multilateral agreements existing in the various organizations and that a first draft Code of Good Practice be prepared by AOAC INTERNATIONAL for consideration at the next IAM. Also, that organizations participating in the IAM be invited to consider the content and utility of such a Code of Good Practice in order to decide at the next IAM whether or not work on such a Code of Good Practice should be continued;
- that the IAM Secretariat be requested to inquire amongst the organizations participating in the IAM whether there is a preliminary interest participating in a comprehensive publication of a compendium of analytical methods of the Codex Alimentarius Commission;
- that the IAM Secretariat be requested to seek comments and prepare a review paper concerning the tasks, utility and future directions of the IAM for consideration at the next IAM; and
- that the IAM, noting the decision by ISO to relinquish the IAM Secretariat, requests CCMAS to recognise the re-assignment of the IAM Secretariat to AOAC INTERNATIONAL and that possible amendments to the Terms of Reference to be proposed by the IAM be considered at the next session of the Committee.

44.   Several delegations, including international organizations were concerned that the report of the IAM would not be appended to the report of the Committee. As the report of the Committee would be widely distributed, the results of the work of the IAM, when appended to the report, would be available to other interested parties, which could not attend either the IAM or session of the Committee to listen to an oral report.

45.   The Committee was informed that due to budgetary constraint, the Secretariat had to reduce its expenditure on publications. Not appending the full report of the IAM to the report of the Session was only one of the methods being taken by the Secretariat to reduce overall cost of publications. This action should not be seen as a reflection on the status of the IAM in the work of the Committee, rather the presence of the Codex Secretariat at the IAM should be seen as a fulfilment of the requirement of

53.   The following remarks were made and agreed during the discussions on Codex General
Methods and methods for Commodity Standards:

(a)   The method for iron in edible oils and fats (IUPAC (1988) 1st Suppl. 2.631, AOAC 990.05)
         should be classified as Type II;

(b)   The term "except edible oils and fats" should be added to the other method for iron (NMKL
         No. 139, 1991);

(c)   The reference to *Pure and Applied Chemistry* should be changed to IUPAC 7th Ed. (1988) 1st
         Suppl. for methods used for oils and fats:

(d)   ISO 8294:1994 (for Cu, Fe, Ni in edible oils) should be added in the list as it was equivalent
         to AOAC 990.05 and ISO 12193:1944 (for Pb in edible oils and fats) as it was equivalent to
         AOAC 994.02;

(e)   As the Guideline Level for aflatoxin in peanuts intended for further processing was currently
         at Step 6, it should be so indicated in the "provision" column; and

(f)   Literature references should be included in notes where other method(s) was (were) referred to
         in the text.

54.   The Codex General Methods for Contaminants along with their assigned Types and the
methods for Commodity Standards considered are attached as Part I and Part II, respectively, of
Appendix IV, together with detailed notes for some of the methods.

55.   The Committee agreed to set up a new *ad hoc* Working Group under the chairmanship of the
Delegation of the USA at its next session.

REVIEW OF METHODS OF ANALYSIS USING OZONE-DEPLETING SUBSTANCES
(Agenda item 12a)

56.   The Committee had for discussion the document[20] prepared by the Representative of AOAC
INTERNATIONAL as a result of discussions[21] held at the 19th Session of the Committee.

57.   The Committee was informed by the Representative of AOAC that, under the Montreal
Protocol of Substances that Deplete the Ozone Layer, production and supply of halogenated
hydrocarbons would be phased out. It was pointed out that there were methods of analysis including
those already endorsed by the Commission which use halogenated hydrocarbons, such as
chlorofluorocarbons and carbon tetrachloride. When these solvents are phased out, there would be a
need for other solvents to replace the ozone depleting solvents. This might affect the status of the
method and replacing a solvent may cause a need for the re-validation of such a method.

58.   The Representative of IDF informed the Committee that a change in solvent in methods
developed by its organization would not require re-numbering of the method. The Representative of
ISO said that if the principle of the method is unchanged re-numbering was unnecessary. The
Representative of AOAC said that replacement of a solvent that affected method performance
necessitated a re-evaluation of the method and hence a new number.

---

[20]   CX/MAS 95/10
[21]   ALINORM 95/23, para 79

59.   The Delegation of Canada said that as the criteria approach was recommended, any method which met the criteria could be used. Furthermore a criterion prohibiting the use of ozone-depleting substances should be considered. The delegation further proposed that it could be included in the criteria that any Type III method using ozone-depleting substances might be withdrawn.

60.   The Delegation of Hungary suggested the investigation of certain methods using micro-volumes of ozone-depleting substances in order to minimize their adverse effect to the environment.

61.   The Committee requested that international organizations working in the field of analysis and sampling identify methods elaborated by them which had been endorsed by the Commission and which use ozone-depleting substances. This information should be communicated to the Codex Secretariat, which should present the identified methods to the *ad hoc* Working Group on Endorsement, based on the information received.

62.   The Codex Committee on Methods of Analysis and Sampling would urge the Commodity Committees to avoid selecting methods of analysis which use ozone-depleting substances.

OTHER BUSINESS AND FUTURE WORK (Agenda item 13)

(I) OTHER BUSINESS

63.   The Committee was informed that the Codex Committee on Milk and Milk Products at its First Session in 1994 decided that in the future it would be appropriate to seek the endorsement of methods of analysis for milk products by the Codex Committee on Methods of Analysis and Sampling (ALINORM 95/11, para. 29). The Executive Committee at its 42nd Session recommended that the Commission make the appropriate changes to the Terms of Reference of the Codex Committee on Methods of Analysis and Sampling to enable it to consider methods of analysis proposed by the Codex Committee on Milk and Milk Products (ALINORM 95/4, para. 37).

64.   On the basis of the above recommendations the Committee agreed to recommend that the Commission amend its Terms of Reference as indicated in Appendix III.

(II) FUTURE WORK

65.   The Committee agreed to continue work on the following items:

- Proposed Draft Codex General Guidelines on Sampling;
- Criteria for evaluating acceptable methods of analysis for Codex purposes;
- Development of objective criteria for assessing the competence of testing laboratories involved in the official import and export control of food;
- Harmonization of reporting of test results corrected for recovery factors;
- Harmonization of analytical terminology in accordance with international standards;
- Report of the IAM on methods of analysis; and
- Endorsement of methods of analysis for Codex purposes.

66.   The Committee agreed to propose that the following new work be undertaken by the Commission:

- Review of methods of analysis using ozone-depleting substances (See paras. 56-62); and

- Measurement uncertainty.

  This new work was proposed by the Delegation of the United Kingdom, who expressed concern that a number of international organizations and accreditation agencies were developing recommendations and requirements regarding measurement uncertainty which were at variance with present practice of the Committee on Methods of Analysis and Sampling. Other delegations requested that consideration of such recommendations and requirements be addressed by the Committee.

67.   Several delegations were desirous that the Commission consider further amendment(s) to the Committee's Terms of Reference to enable it undertake other related work, such as endorsement of microbiological methods to assess safety of food and the development of methods of analysis for the detection of foods produced by biotechnology.

DATE AND PLACE OF THE NEXT SESSION (Agenda Item 14)

68.   The Committee was informed that its 21st Session was tentatively scheduled to be held in Budapest in the 4th week of March 1997, the exact dates will be determined by the Hungarian and the Codex Secretariats.

SUMMARY STATUS OF WORK

SUBJECT	ACTION TO BE TAKEN BY	DOCUMENT REFERENCE ALINORM 97/23
Adoption of the IUPAC/ISO/AOAC Harmonized Guidelines for Internal Quality Control in Analytical Chemistry Laboratories	22nd Session CAC	para. 40 & Appendix II
Amendment to the Committee's Terms of Reference	22nd Session CAC	para. 64 & Appendix III
Proposed Draft Codex General Guidelines on Sampling (At Step 3 of the Procedure)	Codex Secretariat, Governments, CCRVDF, CCFH & 21st Session CCMAS	paras. 10 & 11
Review of Methods of Analysis using Ozone-Depleting Substances	43rd Session Executive Committee, International Organizations, 21st Session CCMAS	paras. 61,62 & 66
Measurement Uncertainty	43rd Session Executive Committee, 21st CCMAS, UK	para. 66
Criteria for Evaluating acceptable Methods of Analysis for Codex purposes	UK & Canada Codex Secretariat 21st Session CCMAS	para. 18
Development of Objective Criteria for Assessing the Competence of Testing Laboratories involved in the Official Import and Export Control of Foods	Finland, CCFICS & 21st Session CCMAS	paras. 22 & 23
Report on the Development of an IUPAC/ISO/AOAC Harmonized Protocol for Recovery Factors	IUPAC 21st session CCMAS	para. 27
Endorsement of Codex Methods and their Classification	21st Session CCMAS	para. 55
Harmonization of Analytical Terminology in accordance with International Standards	USA, Finland, AOAC, ISO, IUPAC, Codex Secretariat, Governments & 21st Session CCMAS	para. 34

LIST OF PARTICIPANTS
LISTE DES PARTICIPANTS

Chairman: Prof. P. Biacs
Président: General Director
                 Central Food Research Institute
                 Herman Ottó út 15
                 H-1022 Budapest, Hungary

Secretary   Dr Váradi, M.
Secrétaire: Scientific Deputy Director
                 Central Food Research Institute
                 1022 Budapest, Herman Ottó u. 15
                 Tel: +36 1 1558 982
                 Fax: +36 1 1558 991

MEMBER COUNTRIES
PAYS MEMBRES

ALBANIA/ALBANIE

Mr Dumani, B.
Chemist-Biologyst
Ministry of Agriculture and Food
Directory of Quality and Inspection of
Food,
Director General, Tirana
Tel/Fax: +355 42 279 24, 279 20
         +355 42 328 97

ALGER/ALGERIE

Ms Chettouf B.
Sous-Directeur
Ministere du Commerce
Palais du Gouvernment,
Alger
73 0051

Mr Khorsi, M
Chef du Département Recherche et
harmonisation des méthodes d'Analyse
Centre Algerien du Control Qualité et
Emballage, CACQE
Ministere du Commerce
2 59 14 36

ARGENTINA/ARGENTINE

Dr Napolitani, C. H.
Chemist
Foods Institute
Estados Unidos 25 -
Buenos Aires 1101

AUSTRALIA/AUSTRALIE

Dr Smith, R. J.
Australian Government Analyst
Australian Government Analytical
Laboratories
P.O.Box 65 Belconnen Act 2616,
Australia
Tel: +61 6 2524923
Fax: +61 6 2524981

AUSTRIA/AUTRICHE

Dr Kapeller, R.
BA für Lebensmitteluntersuchung Linz
A-4021 Linz, Bürgerstr. 47
Tel: +43 732 77907123
Fax: +43 732 77907115

BOLIVIA/BOLIVIE

Mr Mirabal, G.
Food Ingenier
Codex Alimentarius Commission of Bolivia
calle 4 N. 1661 B. Gráficó, La Paz,
Bolivia,

BULGARIA/BULGARIE

Ms Mikouchinska, N.
Medical Doctor,
Expert of Nutrition and Food Control
5, Sreta Nedelia sq., Sofia
Bulgaria,
Tel: 87 52 34
Fax: 80 00 31, 88 37 13

CANADA/CANADA

Dr Lawrence, J. F.
Head,
Food Additives and Contaminants Section,
Food Research Division,
Health Protection Branch,
Health and Welfare Canada
Sir Frederick Banting Building, Ottawa
Ontarion, K1A OL2, Canada
Tel: +1 613 9570946
Fax: +1 613 9414775

Ms Lee, B
Chief, Accreditation and Contaminants
Agriculture + Agrifood Canada
Building No 22., Central Experimental
Farm, Ottawa, Ontario, K1A OC6
Canada
Tel: +1 613 759 1219
Fax: +1 613 759 126

CROATIA/CROATIE

Ms Papic, J
Biochemist
Croatian National Institute of Public Health
Rockefellerova 7, 10000 Zagreb,
Croatia

CZECH REPUBLIC/RÉPUBLIC TCHEQUE

Mr Cuhra, P.
Head of Laboratory
Czech Agricultural and Food Inspection
Pobrezni 10, Prague 8, 186 00,
Czech Republic
Tel: +42 2 24810528
Fax: +42 2 24810528

DENMARK/DANEMARK

Ms. Meyland, I.
Senior Scientist Advisor
National Food Agency
Morkhoj Bygade 19, Dk-2860, Soborg,
Denmark

EGYPT/EGYPTE

Ms Abd El-Kader, A.
Head of Q. C. Sector
Sugar of Integrated Industries Co.
12 Gawad Hosnie, Cairo

FINLAND/FINLANDE

Ms Wallin, H.
Senior Research Scientist
VTT Biotechnology and Food Research
P.O.Box 1500, FIN-02044 VTT,
Finland
Tel: +358 04565193
Fax: +358 04552103

Dr Penttilä, P.-L.
Senior Research Scientist
National Food Administration
P.O.Box 5, FIN-00531 Helsinki,
Finland
Tel: +358 077257621
Fax: +358 077267666

FRANCE/FRANCE

Mr Bourguignon, J. B.
Président de CG d'UMA
Ministere de l'Economie, D.G.C.C.R.F.
59, Bd Vincent Auriol, 75773 Paris cedex
France
Tel: +33 1 44973070
Fax: +33 1 44973038

Ms Normand, N
Responsable Agro-Alimentaire
AFNOR
Tour Europe, 92049 Paris La Defense 2,
France
Tel: +33 1 42915555
Fax: +33 1 42915656

Mr Duran, A.
Inspecteur chargé des questions
d'échantillonnage et de métrologie
Ministere de l'Economie, D.G.C.C.R.F.
59 Bd Vincent Auriol, 75013 Paris, France
Tel: +33 1 44973231
Fax: +33 1 44973037

Ms Janin, F
Directeur de laboratoire
Ministere de l'Agriculture
83 Avenue de St Louis 78000 Versailles,
France

Mr Lestoille, J. P.
Chef de bureau labels et certifications
Ministere de l'Agriculture et de la Peche,
DGAL
175, rue de Chevaleret, 75646 Paris,
France
Tel: +33 1 49555845
Fax: +33 1 49555948

GERMANY/ALLEMAGNE

Dr Palavinskas, R.
BgVV
Thielallee 88-92, 14195 Berlin,
Germany

GREECE/GRECE

Mr Gerakopoulos, D.
Codex Alimentarius Contact Point
Chief, Inspection Service of Agr. Products
Ministry of Agriculture
2 Archarnon Str, 10176 Athens, Greece
Tel: +5246364
Fax: +5243162

HUNGARY/HONGRIE

Ms Boros, I.
Head of Department
Research Institute of Hungarian Sugar
Industry
1084 Budapest, Tolnai L. u. 25,
Hungary
Tel: +36 1 1330 578
Fax: +36 1 1136 418

Ms Bányai, J.
Associate Professor
University of Horticulture and Food
Industry
H-1125 Budapest, Hadik András u. 7,

Dr Domoki, J.
Head of Department
National Institute of Food Hygiene and
Nutrition
1097 Budapest, Gyáli út 3/a,
Hungary
Tel: +36 1 215 4170
Fax: +36 1 215 1545

Dr Gergely, A.
Head of Department
National Institute of Food Hygiene and
Nutrition
1097 Budapest, Gyáli út 3/a
Tel: +36 1 215 4130
Fax: +36 1 215 1545

Dr Kulcsár, F
Officer, Ministry of Agriculture
1051 Budapest, Kossuth tér 11,
Hungary
Tel: +36 1 153 3000

HUNGARY (cont'd)

Dr Matyasovszky, K.
Head of Department
National Institute of Food Hygiene and
Nutrition
1097 Budapest, Gyáli út 3/a, Hungary
Tel: +36 1 215 4130
Fax: +36 1 215 1545

Ms Szerdahelyi, T.
Counsellor/Chemist
Ministry of Agriculture
1050 Budapest, Kossuth tér 11,
Hungary
Tel: +36 1 153 3000

Dr Nagel, V
Main Adviser
National Food Investigation Institute
1095 Budapest, Mester u. 81.
Hungary
Tel: +36 1 215 5440

Dr Tóth-Markus, M
Chemist
Central Food Research Institute
1022 Budapest, Herman o. u. 15.
Tel: +36 1 1558 244
Fax: +36 1 1558 991

INDIA/INDE

Mr Tripathi, J. K
Second Secretary
Embassy of India
14-16 Búzavirág u.
Budapest, 1025
Tel: +36 1 212 3903

INDONESIA/INDONESIE

Dr Dedi Mahdar
Food Technologist
IRDABI
Jalan Ir.H. Juanda No. II. Bogor,
Indonesia
Tel: +62 251 324068; 323339
Fax: +62 251 323339

IRAN/IRAN

Mr Sadrzadeh, P.
Food Analyst
Ministra of Health
Food Control Laboratory (F.O.C.L),
Tehran, 11136
Iran

Mr Rezaeian, M
Food Analyst
Ministry of Health
Food and Drug Control Lab (F.D.C.L),
Tehran, 11136
Iran

JAPAN/JAPON

Mr Hiruta, K.
Chief
Section of Standards and Specifications
Food Sanitation Division
Ministry of Health and Welfare
1-2-2 Kasumigaseki
Chiyoda-Ku
Tokyo, 100-45
Japan

Mr Morita, M
Cnter for Quality Control and Consumer
Service
4-4-1, Konan, Minato-ku, Tokyo 108
Japan

Dr Saito, Y.
Deputy Director-General
National Institute of Health Sciences
1-18-1 Kamiyoga, Setagaya-ku
Tokyo 158, Japan

Dr Uchiyama, S.
Food and Drug Safety Center
HATANO Research Institute
729-5, Ochiai, Hatano-City,
Kanagawa 257, Japan

**KOREA, REPUBLIC OF/
KOREA, RÉPUBLIC DE**

Dr Kim, S.-J.
Senior Research
National Fisheries Research and
Development Agency
468-1, Shirang-ri, Kijang-up, Kijang-gun,
Pusan 619-900
Republic of Korea

Ms Kim, J-H
Analyst
National Fishery Products Inspection Station
103, Wonnan-Dong, Jongro-gu, Seoul, 110-
450
Republic of Korea

LATVIA/LETTONIE

Ms Gratcheva, A
Chief of Laboratory
Central Bread and Bakery Pr. laboratory
Ropaztu 16, LV-1039, Riga,
Latvia

Ms Vakulenko, L.
Technology Engineer
State Inspection of Food Quality
180, Bauskas Str., Riga, Latvija, LV-1076

Ms Kruklite, Z.
Senior Officer
State Inspection of Food Quality
Republikas Lauk 2, Riga, LV 1981,
Latvia

Ms Alaine Ilze
Bacteriologist - Veterinary Surgeon
Riga, Atlasa 7, Latvia

Ms Araja, D
Chief of Laboratory
"Rigas Piensaimnicks"
Riga, Valmieras 2, Latvia

Ms Krûze Maija.
Chief of Laboratory
"Rigas Miesniens"
Riga, Atlasa 7, Latvia

Ms Zviedre, I.
Technology Engineer
Milk Plant Rigas Piena Kombinat
Riga, bauskas str 180, LV-1004

Ms Sukhareva, T.
Chief of Laboratory
Riga, LV-1002, Ventspils, 51
Latvija

Ms Abramova, I
Chief of Laboratory
C/S "TURIBA"
Riga, Latvia
Terbatas str.
Lv 1000

Ms Pastore, E.
Chief of Laboratory
PKS "GRIEZE"
Saldus, Latvia
Saldus galas nomb.

MALAYSIA/MALAISIE

Ms Banaruddin, R.
Admin. Officer (Enforcement)
Porla
1046, 556 Jalan Pevbardaran, 47301 Kelene
Tays

Mr Hashim Man
Administrative Officer (Enforcement)
Porla Pasir Cudang, Bldg. Maritim
LPJ, Tingkat Bawah, 81700 Pasir Gudang,
Malaysia
Tel: +607 2516018
Fax: +607 2510588

Dr Siew Wai Lin
Palm Oil Research Institute of Malaysia
P.O.Box 10620, 50720 Kuala Lumpur
Malaysia

MAROC/MAROQ

Mr Hachimi, L.
Ingenieur, Directeur du laboratoire officiel
d'analyses et du recherches chimiques
25 rue Nichakra Rahal, Casablanca, Maroc
Tel: +302007, 302196
Fax: +301972

THE NETHERLANDS/PAYS-BAS

Mr de Koe, W
Public Health Officer
Ministry of Public Health
Sir Winston Churchill-laan 362, P.O.Box
5406, 2280 H.K. Rijswijk, Netherlands
Tel: +31 70 3406960
Fax: +31 70 3405435

Ms Rentenaar, I.
Senior Standardization Consultant
NNI
P.O.Box 5059 2600 GB Delft,
The Netherlands
Tel: +31 15 690390
Fax: +31 15 390190

Dr H.A.van der Schee
Ministry of Welfare, Health and Culture
Affairs
Regional Inspectorate for Health Protection
Hoogte Kadijk 401
1018 BK Amsterdam
The Netherlands
Tel: +31 20 6237525
Fax: +31 20 6208299

Dr De Ruig, W. G.
State Institute for Quality Control of
Agricultural Products
P.O.Box 230
6700 AE Wageningen
The Netherlands
Tel: +31 317 475474,
      +31 318 417909
Fax: +31 317 417717
      +31 318 417909

NIGER/NIGER (Observer)

Mr Absi, M
Chef Service Hygiene Alimentarie
ONPPC (LANSPEX)
BP 11585, Niamey, Niger

NORWAY/NORVEGE

Ms Nordli, H. S.
Cand. Scient.
Norwegian Food Control
PB.8187 Dep, 0593 Oslo, Norway
Dr Rosness, P. A.
Deputy Director
P.B. 8187 Dep, 0034 Oslo
Norway

PHILIPPINES/PHILIPPINES

Ms Cahanap, A.C.
Chief, Agricultural Chemistry Section,
Bureau of Plant Industry
692 San Andres Street, Malate Manila,
Philippines
Tel: 50 07 08 ; 50 07 79
Fax: 521-76-50

POLAND/POLOGNE

Dr Cwiek-Ludwicka, K.
Food Analyst
National Institute of Hygiene
00-791 Warsaw, 24 Chocimska Str

Dr Jedrzejczak, R.
Head of Lab.
Institute of Agro- and Food Biotechnology
ul. Rakowiecka ; 36
02-532 Warsaw, Poland

Ms Sienkowiec, K.
Laboratory Head
Ministry of Foreign Economic Relations,
Quality Inspection Office
Pilsudskiego 8/12,
81-978 Gdynia, Poland

ROMANIA/ROUMANIE

Mr Alexiu Viorica, G.
Research - Chemist
Food Research Institute
bd Uverturii, Mr 91, Bl, P21, Sc, II, Bp
103

Mr. Spulber, E. G.
Research - Engineur
Food Research Institute
Bd. Banu manta Hz. 1, Bl 1B., Sc A et 6
ap. 27

RUSSIA/RUSSIE

Prof. Skurikhin, I. M.
Head of Laboratory of Food Chemistry
Institute of Nutrition
2/14 Ustinky Proeyd, 109240 Moskow
Russia
Tel: +95 298 38 33
Fax: +95 917 56 72

SINGAPORE/SINGAPOUR

Dr Bloodworth, B. C.
Head (Food Lab)
Institute of Science and Forensic Medicine
NBC Building
Outran Road, Singapore, 0316
Tel: +65 2290724
Fax: +65 2290749

SLOVAKIA/SLOVAQUIE

Mr Dasko, L.
Head of Department
Slovak Agr. and Food Inspection
Mileticova 23, 815 49, Bratislava
Tel: +42 7 211 563
Fax: +42 7 2019280

SLOVENIA/SLOVÉNIE

Ms Marija, B
Ljubljana, Glaverjena 12a
Slovenia

SPAIN/ESPAGNE

Mr Burdaspal, P. A.
Head of Chemical Area
Centro Nacional de Alimentation, Instituto
de Salud Carlos III, Ministerio de Sanindad
I Consumo
28220 - Majadahonda (Madrid)
Tel: +34 1 6381111
Fax: +34 1 6342812

Mr Salas, J
Jefe de Servicio
Centro de Investigation Control de Calidad
C.I.C.C. Avda de Cantabria s/u, 28042
Madrid, Espagña

Dr Vallejo, J. M.
Ingen. Agronomo/Sub.Gen Calidad
Agroalimentaria
Ministerio Agricultura, Pesca y
Alimentacion
P. Infanta Isabel, 1. Madrid-28034
Tel: +341 347 53 94
Fax: +341 347 50 06

SWEDEN/SUEDE

Ms Lönberg, E
Codex Coordinator
National Food Administration
Box 622
S-751 26 Uppsala
Tel: +46 18 175500
Fax: +46 18 105848

Dr Nilsson, A
Manager Quality Assurance
National Food Administration
Box 622
S-751 26 Uppsala
Tel: +46 18 175500
Fax: +4618 105848

SWITZERLAND/SUISSE

Mr Rossier, P.
Chairman of the Swiss National Committee
of Codex Alimentarius
Haslerstrasse 14, CH-3000 Berne 14,
Switzerland

THAILAND/THAILANDE

Ms. Thongtan, N.
Director
Agricultural Chemistry Division,
Department of Agriculture, Ministry of
Agriculture Cooperatives
BKK 10900, Thailand

Ms Syaamananda, C.
Director, Analytical chemistry Laboratory
Thailand Institute of Scientific and
Technological Research
196 Phahonyothin Rd. Chatuchak,
Bangkok, 1900 Thailand

Ms Fasawang, J.
Standards Officer
Office of national Codex Committee,
Ministry of Industry
Thai Industrial Standards Institute, Rama VI
Rd Bangkok 10400, Thailand
Fax: +662 2478741

Ms Pongpituk, V.
State Enterprise Officer, Chemist
Ministry of Science and Technology and
Environment
Thailand Instiute of Scientific and
Technological Research
196 Phahonyothin Rd. Chatuchak, Bangkok
10900, Thailand
Fax: +662 5614771

Mr Srisombai, F
Standards Officer
Department of Foreign Trade, Ministry of
Commerce
Rachadamnern Road, Bldg A.
Panakorndistrict, Bangkok, Thailand, 10500

Ms Wieseswong, U.
Scientist
Department of Foreign Trade, Ministry of
Commerce
BKK 10900, Thailand

Mr Kerdphol, S.
Third Secretary, Royal Thai Embassy,
1025 Budapest, Verecke út 79,
Hungary

Mr Thubthimthai, C
Scientist
Center of Export Inspection and
certification for Agricultural Products
Division of AGricultural Chemistry, Dep.
of Agriculture, BKK 10900,
Ministry of Agricultural Cooperatives
Thailand

Mr Pichalai, A.
Minister Counsellor
Embassy of Thailand
Jozsefhegyi út 28-30, A/B 1025 Budapest,
Hungary

UNITED KINGDOM/ROYAUME UNI

Mr Wood, R.
Head of Department
Ministry of Agriculture, Fisheries and Food
CSL Food Science Laboratory
Norwich Research Park
Colney, Norwich NR4 7UK
Tel: +44 1 603 259350
Fax: +44 1 603 501123

Mr Reynolds, E. B.
Public Analyst
Public Analyst's Laboratory
83, Heavitree Road, Exeter, EX1 2ND
Tel: +44 39272836, +44 39 2434309
Fax: +44 392422691

USA/ETATS UNIS D'AMÉRIQUE

Dr Horwitz, W.
Scientific Adviser
Center for Food Safety and Applied
Nutrition (HFS-500), Food and Drug
Administration
200 "C" Street, S. W. , Washington D.C.
20204 USA
Tel: +1 202 2054346/4046
Fax: +1 202 4017740

Dr Diachenko, G. W.
Director, Division of Product Manufacture
and Use
Center for Food Safety and Applied
Nutrition (HFS-254)
Food and Drug Administration, 200 "C"
Street S. W., Washington, DC 20204 USA
Tel: +1 202 2055320
Fax: +1 202 4017740

Mr Elkins, E.
Chief, Scientist
National Food Procession Assoc.
1401 New York Ave.,
Washington D. C.

Mr Franks, W
Director, Science Division/Ams
US Department of Agriculture
Room 3507, South Building, P.O. Box
96456, Washington,
DC 20090-6456,
USA

Mr McLure, F.
Chief, Statistic Analysis Branch
Food and Drug Administration
200 "C" Street, S. W. Washington DC
20204, USA
Tel: +1 202 2054346
Fax: +1 202+4017740

Dr Rainosek, A. P.
Professor of Statistics
Department of Mathematics and Statistics,
University of South Alabama
Mobile, AL 36688
Tel: +334 460 6264
Fax: +334 460 6166

INTERNATIONAL ORGANIZATIONS ORGANISATIONS INTERNATIONALES

AOAC INTERNATIONAL

Ms Lauwaars, M
European Representative
AOAC International
P.O. Box 153, 6720 AD Bennekom,
Netherlands
Tel: +31 (318) 418 725
Fax: +31 (318) 418 359

Mr Christensen, R.R.
AOAC INTERNATIONAL
Executive Director, General Counsel
2200 Wilson Boulevard, Suite 400,
Arlington, Virginia, USA, 22201-3301
Tel: +1-703-522-3032
Fax: +1-703-522-5468
email: rchristensen@aoac.org

INTERNATIONAL DAIRY FEDERATION (IDF)

Mr Hopkin, E.
Secretary General
IDF
41 Square Vergote, 1040 Brussels Belgium
Tel: +332 733 1690
Fax: +322733 0413

INTERNATIONAL ORGANIZATION FOR STANDARDIZATION (ISO)

Mr Lingner, K.-G.
Deputy Director, Planning and technical Coordination
ISO Central Secretariat
1, rue varembé, CH-1211 Geneva 20, Switzerland
Tel: +41 33 733 34 30

Ms Nagy, E.
Secretary of ISO/TC 34
Hungarian Office for Standardization
Pf. 24., 1450 Budapest 9, Hungary
Tel: +36 1 2183 011
Fax: +36 1 2185 125

Mr Castan, G.
Expert
AFNOR
Tour Europe, 92049 Paris la Defense
Cedex, Paris, France

IUPAC

Dr Parkany, M
ISO Central Secretariat
1, rue Varembé, Geneva, Switzerland
Tel: +41 33 733 34 30

OFFICE INTERNATIONAL DE VIGNE ET DU VIN (OIV)

Ms Mandrou, B.
Professeur
faculté de Pharmacie
F-34060 Montpellier, Cedex 1 (France)

CODEX SECRETARIAT

Mr Baptist, G. O.
Food Standards Officer
Joint FAO/WHO Food Standards Programme, FAO
Via delle Terme di Caracalla, 00100 Rome, Italy
Tel. (0039) 6 5225 3832
Fax. (0039) 6 5225 4593

Dr Yamada, Y.
Food Standards Officer
Joint FAO/WHO Food Standards Programme, FAO
Via delle terme di Caracalla, 00100 Rome, Italy
Tel. (0039) 6 5225 5443
Fax. (0039) 6 5225 4593

Dr Coker, R.
Principal Natural Products Scientist
Natural Resources Institute
Central Avenue, Chatham Maritine, Chatham, Kent, MR4 4TB, UK
Tel: +44 1634 883455
Fax: +44 1634 880066

RECOMMENDED HARMONIZED GUIDELINES FOR INTERNAL QUALITY CONTROL
IN ANALYTICAL CHEMISTRY LABORATORIES
(At Step 8 of the Procedure)

The following document is recommended for adoption for Codex purposes by the 22nd Session of the Commission.

Harmonized Guidelines for Internal Quality Control in Analytical Chemistry Laboratories (Pure and Appl. Chem., Vol. 67, No. 4, pp. 649-666, 1995).

AMENDMENT OF THE TERMS OF REFERENCE OF THE
CODEX COMMITTEE ON METHODS OF ANALYSIS AND SAMPLING
(Submitted to the Commission for adoption)[1]

Amend paragraph (d) of the Terms of Reference of the Committee (*Codex Alimentarius Procedural Manual*, Eighth Edition, page 133) as follows (struck-out text to be deleted and italicized text to be inserted):

(d)   to consider, amend, if necessary, and endorse, as appropriate, methods of analysis and sampling proposed by Codex (Commodity) Committees, except that methods of analysis and sampling for residues of pesticides or veterinary drugs in food, the assessment of micro-biological quality and safety in food, *and* the assessment of specifications for food additives, do not fall within the terms of reference of this Committee.

---

[1]   ALINORM 95/4, para. 37.

**LIST OF METHODS OF ANALYSIS CONSIDERED
BY THE TWENTIETH SESSION OF THE CODEX COMMITTEE ON METHOD OF
ANALYSIS AND SAMPLING**

Part I: Codex General Methods for Contaminants

Part II: Methods of Analysis for Commodity Standards

Notes on Parts I and II.

PART I - CODEX GENERAL METHODS FOR CONTAMINANTS

PROVISION	METHOD	PRINCIPLE	TYPE
Cadmium	AOAC 982.23	Anodic stripping voltammetry	III
Cadmium	NMKL No. 139, 1991	Atomic absorption spectrometry	III
Chromium	NMKL No. 139, 1991	Atomic absorption spectrometry	II
Copper (in edible oils and fats)	IUPAC 7th ed. (1988) 1st Suppl. 2.631 AOAC 990.05 ISO 8294:1994	Direct graphite furnace atomic absorption spectrometry	II
Iron (in edible oils and fats)	IUPAC 7th ed. (1988) 1st Suppl. 2.631 AOAC 990.05 ISO 8294:1994	Direct graphite furnace atomic absorption spectrometry	II
Iron (except in edible oils and fats)	NMKL No. 139, 1991	Atomic absorption spectrometry	II
Lead	AOAC 982.23	Anodic stripping voltammetry	III
Lead (in edible oils and fats)	IUPAC 7th ed. (1988) 1st Suppl. 2.632 AOAC 994.02 ISO 12193:1994	Direct graphite furnace atomic absorpton spectrometry	II
Lead	NMKL No. 139, 1991	Atomic absorption spectrometry	III
Nickel (in edible oils and fats)	IUPAC 7th ed. (1988) 1st Suppl. 2.631 AOAC 990.05 ISO 8294:1994	Direct graphite furnace atomic absorption spectrometry	II
Tin (in canned foods)	AOAC 985.16	Atomic absorption spectrometry	III
Zinc	NMKL No. 139, 1991	Atomic absorption spectrometry	III

PART II - METHODS OF ANALYSIS FOR COMMODITY STANDARDS

Page 1

Serial No	Commodity Standard No.	Provision	Method	Principle	Type	Status
249	Special foods 980	Copper, manganese, zinc, magnesium, iron Cu: >60 mg, Mn: >5 µg, Zn: >0.5 mg. Mg: >6 mg and Fe: >0.15 mg/100 kcal	AOAC 984.27	ICP emission spectrometry		NE
251	Foods with low-sodium content (including salt substitutes) 053-1981	Sodium and potassium Na: < 120 mg/100 g, K: No limit	AOAC 984.27	ICP emission spectrometry		NE
252	Foods with low-sodium content (including salt substitutes) 053-1981	Calcium and magnesium Mg: < 20 % of sum of potassium, calcium, ammonium cations	AOAC 965.09	Atomic absorption spectrophotometry		NE
253	Foods with low-sodium content (including salt substitutes) 053-1981	Ammonium < 3 % (m/m)	AOAC 920.03	Magnesium oxide		NE
254	Foods with low-sodium content (including salt substitutes) 053-1981	Phosphorous < 4 % (m/m)	AOAC 984.27	ICP emission spectrometry		NE
580	Guidelines for nutrition labelling CAC/GL 2-1985	Polyunsaturated fat	AOCS Ce 1c-89	Gas liquid chromatography	IV	TE
581	Guidelines for nutrition labelling CAC/GL 2-1985	saturated fat	AOCS Ce 1c-89	Gas liquid chromatography	IV	TE
634	Quick frozen fish sticks (fish fingers) Fish portions & fish fillets - breaded or in batter 166-1989	Histamine 10 mg/100 g	AOAC 977.13	Fluorimetry	II	E
635	Infant formula and follow-up formula 72-1981 & 156-1987	Total dietary fibre	J. Publ. Analysts (1993) 29 (2)	Englyst method		NE

PART II - METHODS OF ANALYSIS FOR COMMODITY STANDARDS

Page 2

Serial No	Commodity Standard No.	Provision	Method	Principle	Type	Status
636	Quick frozen fish sticks (Fish fingers) Fish portions & fish fillets - breaded or in Batter 166-1989	Fish core	AOAC 971.13	Immersion and weighing	I	E
637	Milk	Aflatoxin M1 0.05 µg/kg	IDF STD. 171:1995	Immunoaffinity column & LC	II	E
638	Milk & dried milk A-5 (milk powder)	Aflatoxin M1 0.05 µg/kg	IDF Std. 111A:1990	TLC/LC		NE
639	Fluid milk	Aflatoxin M1 0.05 µg/kg	AOAC 986.16	HPLC		NE
640	Peanuts (intended for further processing)	Aflatoxin, total 15 µg/kg (Step 6)	AOAC 975.36	Romer mini colmn	III	E
641	Peanuts (intended for further processing)	Aflatoxin, total 15 µg/kg (Step 6)	AOAC 979.18	Holaday-Velasco mini column	III	E
642	Corn	Aflatoxin, total	AOAC 979.18	Holaday-Velasco mini column	II	E
643	Peanuts	Aflatoxin, total 15 µg/kg (Step 6)	AOAC 990.34	ELISA		NE
644	Peanuts & peanut products	Aflatoxin, total 15 µg/kg (Step 6)	AOAC 968.22	CB Method		NE
645	Peanuts & peanut products	Aflatoxin, total 15 µg/kg (Step 6)	AOAC 970.45	BF method		NE
646	Peanuts (Raw)	Aflatoxin, total 15 µg/kg (Step 6)	AOAC 993.17	TLC	III	E
647	Peanuts (Raw)	Aflatoxin, total 15 µg/kg (Step 6)	AOAC 991.31	Immunoaffinity column (Aflatest)	II	E

PART II - METHODS OF ANALYSIS FOR COMMODITY STANDARDS

Page 3

Serial No.	Commodity Standard No.	Provision	Method	Principle	Type	Status
648	Corn	Aflatoxin, total	AOAC 990.34	ELISA		NE
649	Cotton Seed	Aflatoxin, total	AOAC 990.34	ELISA		NE

NOTES

Part I   Codex General Methods for Contaminants

Cadmium: The WG considered that there was already a Type II Codex general method for Cadmium and so classsified the two new methods (AOAC 982.23 and NMKL 139) as Type III.

Copper: This method (AOAC 990.05) had already been classified as Type II for fats and oils. It was recommended that the Codex Committee on Methods of Analysis and Sampling agreed to change the colorimetric method (AOAC 960.40), which had been classified as Type II for fats and oils to Type III, in order to avoid having more than one Type II method for fats and oils. The Committee noted that ISO 8294:1994 is identical to the IUPAC method. It was also proposed that the appropriate IUPAC numbering be used for references to IUPAC methods.

Iron: The WG recommended that the IUPAC method and AOAC 990.05 be classified as Type II for fats and oils and the Atomic absorption method (NMKL No. 139), be classified as Type II.

Lead: The IUPAC method had already been classified as Type II for fats and oils. The WG noted that AOAC 994.02 and ISO 12193:1994 were equivalent methods. It was also observed that there was already in place a colorimetric dithizone method (AOAC 934.07) for lead in fats and oils. The WG therefore proposed that the Committee on Fats and Oils should consider deleting the method because the method is not sensitive enough to detect lead at the specification level. If however, the Commodity Committee would rather retain the method, it should be classified as Type III.

Nickel: The IUPAC method and AOAC 990.05 were recommended as Type II for fats and oils.

Tin: The method (AOAC 985.16) had previously been classified as Type III for a canned food, therefore the WG retained this classification.

Zinc: The method (NML No. 139) was classified as Type III.

Part II   Methods of Analysis for Commodity Standards

The following comments were made:

66-128, 689-861   The methods for Sugars and Fats and Oils respectively, were not considered because the respective Commodity Committees were in the process of considering comments to circular letters which were circulated. The Working Group therefore recommended that consideration of these methods be suspended, pending the results of the actions taken by the respective Commodity Committees. The Working Group urged the Codex Committee on Methods of Analysis and Sampling to request its members to provide comments on CL 1995/22-FO directly to the Codex Committee on Fats and Oils.

138   The Secretariat was requested to contact the secretariat of AIIBP to obtain the necessary information, regarding the applicability of the method.

249, 251-254 It was noted that these methods have not been collaboratively studied for these commodities which contain salt substitutes and that there were no methods applicable to these matrices that meet the criteria of the Codex Committee on Methods of Analysis and Sampling. In view of this, the WG recommended the withdrawal of the temporary endorsement earlier granted to the reference methods and their deletion from being considered for endorsement.

353, 354, 475 & 489 The temporary endorsements were retained. The Secretariat was requested to bring the status and previous concerns of the Codex Committee on Methods of Analysis and Sampling to the attention of the Codex Committee on Natural Mineral Water.

435 & 503: The WG recommended that the Secretariat of the Codex Committee on Cereals, Pulses and Legumes be contacted to consider the comments earlier made and request their recommendations or concurrence with the proposals of the WG. If action was not taken the WG would recommend the withdrawal of the temporary endorsement.

509: Same recommendation as for 435 & 503, except that the Secretariat should contact the Codex Committee on Processed Fruits and Vegetables.

580 & 581 The WG temporarily endorsed these methods as Type IV and requested the Secretariat to contact the American Oil Chemists Society for method validation information which if available and found to be satisfactory, would enable the WG to recommend full endorsement.

635 The WG observed that there was already a Type I method (AOAC 991.43) for the determination of dietary fibre. The request by the Delegation of the United Kingdom for the WG to consider the Englyst method was not supported because the WG observed that the method and indeed the already endorsed one, could not determine the specified level for carrageenan and there was no specification for the level of the of carry-over fiber.

638 & 639 The WG did not recommend endorsement of these methods because, the WG noted that the IDF methods could not detect down to the limits prescribed. It was noted that this information should be transmitted to the IDF/ISO/AOAC Tripartite Working Group on Methods of Analysis, which recommend methods for milk products to the Codex Alimentarius Commission through the Codex Committee on Milk and Milk Products.

640, 641, 646 & 647 The WG recommended Type III classification for all except 647 which it classified as Type II. It was also noted that CEN had specified the size of the column and the method was no longer a proprietary one.

642 Recommended for endorsement as Type II method since it can measure levels higher than 10 μg/kg which is adequate for the guideline level.

643, 644, 645 & 648 All were not recommended for endorsement because they are not sensitive enough for analyses at the guideline levels. It was also noted that 648 was a proprietary method and one of the solvents used in 644 is chloroform - an ozone-depleting substance.

649 Cottonseed, not being for direct human consumption as food, the WG did not find this reference appropriate for consideration and so recommended its deletion.

13   ENVIRONMENT AND HEALTH PROTECTION. SAFETY

13.020   Environment protection in general

ISO/DIS Guide 64   *Ed. 1   6 p.*   *TC 207*
Guide for the inclusion of environmental aspects in product
standards

ISO/DIS 14001   *Ed. 1   17 p.*   *TC 207 / SC 1*
Environmental management systems — Specification with
guidance for use

ISO/DIS 14004   *Ed. 1   42 p.*   *TC 207 / SC 1*
Environmental management systems — General guidelines on
principles, systems and supporting techniques

ISO/DIS 14010   *Ed. 1   6 p.*   *TC 207 / SC 2*
Guidelines for environmental auditing — General principles

ISO/DIS 14011   *Ed. 1   9 p.*   *TC 207 / SC 2*
Guidelines for environmental auditing — Audit
procedures — Auditing of environmental management systems

ISO/DIS 14012   *Ed. 1   7 p.*   *TC 207 / SC 2*
Guidelines for environmental auditing — Qualification criteria for
environmental auditors

13.030   Solid wastes

ISO 6961:1982   *Ed. 1   6 p. (C)*   *TC 85 / SC 5*
Long-term leach testing of solidified radioactive waste forms

ISO 6962:1982   *Ed. 1   8 p. (D)*   *TC 85 / SC 5*
Standard method for testing the long term alpha irradiation
stability of solidified high-level radioactive waste forms

ISO/DIS 11932   *Ed. 1*   *TC 85*
Activity measurements of solid materials considered for recycling,
re-use or disposal as non-radioactive waste

13.040   Air quality

13.040.10   General aspects

ISO 4225:1994   *Ed. 2   12 p. (F)*   *TC 146 / SC 4*
Air quality — General aspects — Vocabulary
Bilingual edition

ISO 4226:1993   *Ed. 2   2 p. (A)*   *TC 146 / SC 4*
Air quality — General aspects — Units of measurement

ISO 6879:1995   *Ed. 2   6 p. (C)*   *TC 146 / SC 4*
Air quality — Performance characteristics and related concepts
for air quality measuring methods

ISO 7708:1995   *Ed. 1   9 p. (E)*   *TC 146 / SC 2*
Air quality — Particle size fraction definitions for health-related
sampling

ISO 8756:1994   *Ed. 1   4 p. (B)*   *TC 146 / SC 4*
Air quality — Handling of temperature, pressure and humidity
data

ISO 9169:1994   *Ed. 1   18 p. (J)*   *TC 146 / SC 4*
Air quality — Determination of performance characteristics of
measurement methods

ISO/DIS 11222   *Ed. 1   8 p.*   *TC 146 / SC 4*
Air quality — Comparison of mean values determined from air
quality measurements with prescribed values

ISO/DIS 13752   *Ed. 1   17 p.*   *TC 146 / SC 4*
Air quality — Assessment of the uncertainty of a measurement
method under field conditions using a second method as a
reference

13.040.20   Ambient atmospheres

ISO 4219:1979   *Ed. 1   4 p. (B)*   *TC 146 / SC 3*
Air quality — Determination of gaseous sulphur compounds in
ambient air — Sampling equipment

ISO 4220:1983   *Ed. 1   5 p. (C)*   *TC 146 / SC 3*
Ambient air — Determination of a gaseous acid air pollution
index — Titrimetric method with indicator or potentiometric
end-point detection

ISO 4221:1980   *Ed. 1   7 p. (D)*   *TC 146 / SC 3*
Air quality — Determination of mass concentration of sulphur
dioxide in ambient air — Thorin spectrophotometric method

ISO/DIS 4222   *Ed. 1*   *TC 146 / SC 3*
Ambient air — Measurement of particulate fall-out — Horizontal
deposit gauge method

ISO/DIS 4224   *Ed. 1   16 p.*   *TC 146 / SC 3*
Ambient air — Determination of carbon
monoxide — Non-dispersive infrared spectrometric method

ISO/TR 4227:1989   *Ed. 1   15 p. (H)*   *TC 146 / SC 4*
Planning of ambient air quality monitoring

ISO 6767:1990   *Ed. 1   31 p. (E)*   *TC 146 / SC 3*
Ambient air — Determination of the mass concentration of sulfur
dioxide — Tetrachloromercurate (TCM)/pararosaniline method

ISO 6768:1985   *Ed. 1   12 p. (F)*   *TC 146 / SC 3*
Ambient air — Determination of the mass concentration of
nitrogen dioxide — Modified Griess-Saltzman method

13.040.20

ISO/DIS 6768   *Ed. 2*   *11 p.*   *TC 146 / SC 3*
Ambient air — Determination of mass concentration of nitrogen
dioxide — Modified Griess-Saltzman method (Revision of ISO
6768:1985)

ISO 7168:1985   *Ed. 1*   *8 p. (D)*   *TC 146 / SC 4*
Air quality — Presentation of ambient air quality data in
alphanumerical form

ISO 7996:1985   *Ed. 1*   *9 p. (E)*   *TC 146 / SC 3*
Ambient air — Determination of the mass concentration of
nitrogen oxides — Chemiluminescence method

ISO 8186:1989   *Ed. 1*   *8 p. (D)*   *TC 146 / SC 3*
Ambient air — Determination of the mass concentration of
carbon monoxide — Gas chromatographic method

ISO 9359:1989   *Ed. 1*   *12 p. (F)*   *TC 146 / SC 4*
Air quality — Stratified sampling method for assessment of
ambient air quality

ISO 9835:1993   *Ed. 1*   *9 p. (E)*   *TC 146 / SC 3*
Ambient air — Determination of a black smoke index

ISO 9855:1993   *Ed. 1*   *7 p. (D)*   *TC 146 / SC 3*
Ambient air — Determination of the particulate lead content of
aerosols collected on filters — Atomic absorption spectrometric
method

ISO 10312:1995   *Ed. 1*   *51 p. (U)*   *TC 146 / SC 3*
Ambient air — Determination of asbestos fibres — Direct
transfer transmission electron microscopy method

ISO 10313:1993   *Ed. 1*   *9 p. (E)*   *TC 146 / SC 3*
Ambient air — Determination of the mass concentration of
ozone — Chemiluminescence method

ISO/DIS 10498   *Ed. 1*   *7 p.*   *TC 146 / SC 3*
Ambient air — Determination of sulfur dioxide — Ultraviolet
fluorescence method

ISO/DIS 10529   *Ed. 1*   *26 p.*   *TC 146 / SC 3*
Ambient air — Determination of the mass concentration of
gaseous and soluble particulate fluorine-containing
compounds — Method by filter sampling and ion-selective
electrode analysis

ISO/DIS 11454   *Ed. 1*   *7 p.*   *TC 126*
Tobacco and tobacco products — Determination of vapour-phase
nicotine in air — Gas-chromatographic method

ISO/DIS 13794   *Ed. 1*   *100 p.*   *TC 146 / SC 3*
Ambient air — Determination of asbestos
fibres — Indirect-transfer transmission electron microscopy
method

13.040.30 Workplace atmospheres

ISO 8518:1990   *Ed. 1*   *10 p. (E)*   *TC 146 / SC 2*
Workplace air — Determination of particulate lead and lead
compounds — Flame atomic absorption spectrometric method

ISO 8672:1993   *Ed. 1*   *25 p. (M)*   *TC 146 / SC 2*
Air quality — Determination of the number concentration of
airborne inorganic fibres by phase contrast optical
microscopy — Membrane filter method

ISO 8760:1990   *Ed. 1*   *6 p. (C)*   *TC 146 / SC 2*
Work-place air — Determination of mass concentration of carbon
monoxide — Method using detector tubes for short-term
sampling with direct indication

ISO 8761:1989   *Ed. 1*   *8 p. (D)*   *TC 146 / SC 2*
Work-place air — Determination of mass concentration of
nitrogen dioxide — Method using detector tubes for short-term
sampling with direct indication

ISO 8762:1988   *Ed. 1*   *9 p. (E)*   *TC 146 / SC 2*
Workplace air — Determination of vinyl chloride — Charcoal
tube/gas chromatographic method

ISO 9486:1991   *Ed. 1*   *12 p. (F)*   *TC 146 / SC 2*
Workplace air — Determination of vaporous chlorinated
hydrocarbons — Charcoal tube/solvent desorption/gas
chromatographic method

ISO 9487:1991   *Ed. 1*   *12 p. (F)*   *TC 146 / SC 2*
Workplace air — Determination of vaporous aromatic
hydrocarbons — Charcoal tube/solvent desorption/gas
chromatographic method

ISO/DIS 9976   *Ed. 1*   *15 p.*   *TC 146 / SC 2*
Workplace air — Determination of concentrations of C3 to C10
hydrocarbons — Sorbent tube/thermal desorption/capillary gas
chromatographic method

ISO/DIS 9977   *Ed. 1*   *11 p.*   *TC 146 / SC 2*
Workplace air — Determination of acrylonitrile — Pumped
sorbent tube/thermal desorption/gas chromatographic method

ISO/DIS 11041   *Ed. 1*   *20 p.*   *TC 146 / SC 2*
Workplace air — Determination of particulate arsenic, arsenic
compounds and arsenic trioxide vapour — Method by hydride
generation and atomic absorption spectrometry

ISO/DIS 11174   *Ed. 1*   *15 p.*   *TC 146 / SC 2*
Workplace air — Determination of particulate cadmium and
cadmium compounds — Flame and electrothermal atomic
absorption spectrometric method

13.040.40 Stationary source emissions

ISO 7934:1989   *Ed. 1*   *6 p. (C)*   *TC 146 / SC 1*
Stationary source emissions — Determination of the mass
concentration of sulfur dioxide — Hydrogen peroxide/barium
perchlorate/Thorin method

ISO 7935:1992   *Ed. 1*   *11 p. (F)*   *TC 146 / SC 1*
Stationary source emissions — Determination of the mass
concentration of sulfur dioxide — Performance characteristics of
automated measuring methods

ISO 9096:1992   *Ed. 1*   *30 p. (P)*   *TC 146 / SC 1*
Stationary source emissions — Determination of concentration
and mass flow rate of particulate material in gas-carrying
ducts — Manual gravimetric method

ISO 10155:1995   *Ed. 1*   *18 p. (J)*   *TC 146 / SC 1*
Stationary source emissions — Automated monitoring of mass
concentrations of particles — Performance characteristics, test
methods and specifications

ISO 10396:1993   *Ed. 1*   *13 p. (G)*   *TC 146 / SC 1*
Stationary source emissions — Sampling for the automated
determination of gas concentrations

ISO 10397:1993   *Ed. 1*   *19 p. (K)*   *TC 146 / SC 1*
Stationary source emissions — Determination of asbestos plant
emissions — Method by fibre count measurement

ISO 10780:1994   *Ed. 1*   *19 p. (K)*   *TC 146 / SC 1*
Stationary source emissions — Measurement of velocity and
volume flowrate of gas streams in ducts

ISO/DIS 10849   *Ed. 1*   *19 p.*   *TC 146 / SC 1*
Stationary source emissions — Determination of the mass
concentration of nitrogen oxides — Performance characteristics
of automated measuring systems

ISO/DIS 11042-1   *Ed. 1*   *TC 192*
Gas turbines — Exhaust gas emission —
Part 1: Measurement and evaluation

ISO/DIS 11042-2   *Ed. 1*   *33 p.*   *TC 192*
Gas turbines — Exhaust gas emission —
Part 2: Automated emission monitoring

13.040.50 Transport exhaust emissions

ISO 789-4:1986   *Ed. 2   7 p. (D)*   *TC 23 / SC 2*
Agricultural tractors — Test procedures —
Part 4: Measurement of exhaust smoke

ISO 3173:1974   *Ed. 1   26 p. (M)*   *TC 22 / SC 5*
Road vehicles — Apparatus for measurement of the opacity of
exhaust gas from diesel engines operating under steady state
conditions

ISO 3929:1995   *Ed. 2   6 p. (C)*   *TC 22 / SC 5*
Road vehicles — Measurement methods for exhaust gas
emissions produced during inspection or maintenance

ISO 3930:1993   *Ed. 2   16 p. (H)*   *TC 22 / SC 5*
Road vehicles — Measurement equipment for exhaust gas
emissions during inspection or maintenance — Technical
specifications

ISO/TR 4011:1976   *Ed. 1   17 p. (J)*   *TC 22 / SC 5*
Road vehicles — Apparatus for measurement of the opacity of
exhaust gas from diesel engines

ISO 6460:1981   *Ed. 1   14 p. (G)*   *TC 22 / SC 22*
Road vehicles — Measurement method of gaseous pollutants
emitted by motorcycles equipped with a controlled ignition engine

ISO 6855:1983   *Ed. 2   13 p. (G)*   *TC 22 / SC 23*
Road vehicles — Measurement methods for gaseous pollutants
emitted by mopeds equipped with a controlled ignition engine

ISO 6970:1994   *Ed. 1   9 p. (E)*   *TC 22 / SC 23*
Motorcycles and mopeds — Pollution tests — Chassis
dynamometer bench

ISO 7644:1988   *Ed. 1   3 p. (B)*   *TC 22 / SC 5*
Road vehicles — Measurement of opacity of exhaust gas from
compression-ignition (diesel) engines — Lug-down test

ISO 7645:1988   *Ed. 1   2 p. (A)*   *TC 22 / SC 5*
Road vehicles — Measurement of opacity of exhaust gas from
compression-ignition (diesel) engines — Steady single-speed test

ISO/DIS 8178-1   *Ed. 1*   *TC 70 / SC 8*
Reciprocating internal combustion engines — Exhaust emission
measurement —
Part 1: Test-bed measurement of gaseous and particulate
exhaust emissions

ISO/DIS 8178-2   *Ed. 1   22 p.*   *TC 70 / SC 8*
Reciprocating internal combustion engines — Exhaust emission
measurement —
Part 2: Measurement of gaseous and particulate exhaust
emissions at site

ISO 8178-3:1994   *Ed. 1   6 p. (C)*   *TC 70 / SC 8*
Reciprocating internal combustion engines — Exhaust emission
measurement —
Part 3: Definitions and methods of measurement of exhaust gas
smoke under steady-state conditions

ISO/DIS 8178-4   *Ed. 1   17 p.*   *TC 70 / SC 8*
Reciprocating internal combustion engines — Exhaust emission
measurement —
Part 4: Test cycles for different engine applications

ISO/DIS 8178-5   *Ed. 1*   *TC 70 / SC 8*
Reciprocating internal combustion engines — Exhaust emission
measurement —
Part 5: Test fuels

ISO/DIS 8178-6   *Ed. 1   17 p.*   *TC 70 / SC 8*
Reciprocating internal combustion engines — Exhaust emission
measurement —
Part 6: Test report

ISO/DIS 8178-7   *Ed. 1*   *TC 70 / SC 8*
Reciprocating internal combustion engines — Exhaust emission
measurement —
Part 7: Engine family

ISO/DIS 8178-8   *Ed. 1*   *TC 70 / SC 8*
Reciprocating internal combustion engines — Exhaust emission
measurement —
Part 8: Engine group determination

ISO/TR 9310:1987   *Ed. 1   14 p. (G)*   *TC 22 / SC 5*
Road vehicles — Smoke measurement of compression-ignition
(diesel) engines — Survey of short in-service tests

ISO/DIS 10054   *Ed. 1   16 p.*   *TC 22 / SC 5*
Internal combustion compression-ignition
engines — Measurement apparatus for smoke from engines
operating under steady-state conditions — Filter-type
smokemeter

ISO/DIS 11042-1   *Ed. 1*   *TC 192*
Gas turbines — Exhaust gas emission —
Part 1: Measurement and evaluation

ISO/DIS 11042-2   *Ed. 1   33 p.*   *TC 192*
Gas turbines — Exhaust gas emission —
Part 2: Automated emission monitoring

ISO/DIS 11614   *Ed. 1   63 p.*   *TC 22 / SC 5*
Reciprocating internal combustion compression-ignition
engines — Apparatus for measurement of the opacity and for
determination of the light absorption coefficient of exhaust gas
(/Combination of ISO 3173-1974 and ISO/TR 4011-1976)

13.060   Water quality

ISO 6107-1:1996   *Ed. 3   29 p. (N)*   *TC 147 / SC 1*
Water quality — Vocabulary —
Part 1
Trilingual edition

ISO 6107-2:1989   *Ed. 2   31 p. (P)*   *TC 147 / SC 1*
Water quality — Vocabulary —
Part 2
Trilingual edition

ISO 6107-3:1993   *Ed. 2   22 p. (L)*   *TC 147 / SC 1*
Water quality — Vocabulary —
Part 3
Trilingual edition

ISO 6107-4:1993   *Ed. 2   8 p. (D)*   *TC 147 / SC 1*
Water quality — Vocabulary —
Part 4
Quadrilingual edition

ISO 6107-5:1996   *Ed. 2   20 p. (K)*   *TC 147 / SC 1*
Water quality — Vocabulary —
Part 5
Trilingual edition

ISO 6107-6:1996   *Ed. 2   18 p. (J)*   *TC 147 / SC 1*
Water quality — Vocabulary —
Part 6
Trilingual edition

ISO 6107-7:1990   *Ed. 1   16 p. (H)*   *TC 147 / SC 1*
Water quality — Vocabulary —.
Part 7
Trilingual edition

ISO 6107-8:1993   *Ed. 1   16 p. (H)*   *TC 147 / SC 1*
Water quality — Vocabulary —
Part 8
Quadrilingual edition

ISO/DIS 6107-9   *Ed. 1   32 p.*   *TC 147 / SC 1*
Water quality — Vocabulary —
Part 9: Complete list of terms in ISO 6107
Bilingual edition

13.060.30

13.060.30 Sewage water disposal and treatment

ISO 8099:1985   *Ed. 1   3 p. (B)*   *TC 188*
Small craft — Toilet retention and recirculating systems for the
treatment of toilet waste

ISO/DIS 8099-1   *Ed. 1   3 p.*   *TC 188*
Small craft — Waste water retention and treatment —
Part 1: Toilet retention systems (Revision of ISO 8099:1985)

13.060.40 Examination of water, waste water and sludge

ISO 5663:1984   *Ed. 1   4 p. (B)*   *TC 147 / SC 2*
Water quality — Determination of Kjeldahl nitrogen — Method
after mineralization with selenium

ISO 5664:1984   *Ed. 1   3 p. (B)*   *TC 147 / SC 2*
Water quality — Determination of ammonium — Distillation and
titration method

ISO 5666-1:1983   *Ed. 1   6 p. (C)*   *TC 147 / SC 2*
Water quality — Determination of total mercury by flameless
atomic absorption spectrometry —
Part 1: Method after digestion with permanganate-peroxodisulfate

ISO 5666-2:1983   *Ed. 1   8 p. (D)*   *TC 147 / SC 2*
Water quality — Determination of total mercury by flameless
atomic absorption spectrometry —
Part 2: Method after pretreatment with ultraviolet radiation

ISO 5666-3:1984   *Ed. 1   6 p. (C)*   *TC 147 / SC 2*
Water quality — Determination of total mercury by flameless
atomic absorption spectrometry —
Part 3: Method after digestion with bromine

ISO 5667-1:1980   *Ed. 1   13 p. (G)*   *TC 147 / SC 6*
Water quality — Sampling —
Part 1: Guidance on the design of sampling programmes

    Technical Corrigendum 1:1996 to ISO 5667-1:1980
               Ed. 1   1 p.   *TC 147 / SC 6*

ISO 5667-2:1991   *Ed. 2   9 p. (E)*   *TC 147 / SC 6*
Water quality — Sampling —
Part 2: Guidance on sampling techniques

ISO 5667-3:1994   *Ed. 2   31 p. (M)*   *TC 147 / SC 6*
Water quality — Sampling —
Part 3: Guidance on the preservation and handling of samples

ISO 5667-4:1987   *Ed. 1   5 p. (C)*   *TC 147 / SC 6*
Water quality — Sampling —
Part 4: Guidance on sampling from lakes, natural and man-made

ISO 5667-5:1991   *Ed. 1   8 p. (D)*   *TC 147 / SC 6*
Water quality — Sampling —
Part 5: Guidance on sampling of drinking water and water used
for food and beverage processing

ISO 5667-6:1990   *Ed. 1   9 p. (E)*   *TC 147 / SC 6*
Water quality — Sampling —
Part 6: Guidance on sampling of rivers and streams

ISO 5667-7:1993   *Ed. 1   16 p. (H)*   *TC 147 / SC 6*
Water quality — Sampling —
Part 7: Guidance on sampling of water and steam in boiler plants

ISO 5667-8:1993   *Ed. 1   9 p. (E)*   *TC 147 / SC 6*
Water quality — Sampling —
Part 8: Guidance on the sampling of wet deposition

ISO 5667-9:1992   *Ed. 1   8 p. (D)*   *TC 147 / SC 6*
Water quality — Sampling —
Part 9: Guidance on sampling from marine waters

ISO 5667-10:1992   *Ed. 1   10 p. (E)*   *TC 147 / SC 6*
Water quality — Sampling —
Part 10: Guidance on sampling of waste waters

ISO 5667-11:1993   *Ed. 1   10 p. (E)*   *TC 147 / SC 6*
Water quality — Sampling —
Part 11: Guidance on sampling of groundwaters

ISO 5667-12:1995   *Ed. 1   34 p. (Q)*   *TC 147 / SC 6*
Water quality — Sampling —
Part 12: Guidance on sampling of bottom sediments

ISO/DIS 5667-13   *Ed. 1   28 p.*   *TC 147 / SC 6*
Water quality — Sampling —
Part 13: Guidance on sampling of sewage, waterworks and
related sludges

ISO 5813:1983   *Ed. 1   5 p. (C)*   *TC 147 / SC 2*
Water quality — Determination of dissolved
oxygen — Iodometric method

ISO 5814:1990   *Ed. 2   8 p. (D)*   *TC 147 / SC 2*
Water quality — Determination of dissolved
oxygen — Electrochemical probe method

ISO 5815:1989   *Ed. 2   6 p. (C)*   *TC 147 / SC 2*
Water quality — Determination of biochemical oxygen demand
after 5 days (BOD 5) — Dilution and seeding method

ISO 5961:1994   *Ed. 2   10 p. (E)*   *TC 147 / SC 2*
Water quality — Determination of cadmium by atomic absorption
spectrometry

ISO 6058:1984   *Ed. 1   3 p. (B)*   *TC 147 / SC 2*
Water quality — Determination of calcium content — EDTA
titrimetric method

ISO 6059:1984   *Ed. 1   4 p. (B)*   *TC 147 / SC 2*
Water quality — Determination of the sum of calcium and
magnesium — EDTA titrimetric method

ISO 6060:1989   *Ed. 2   4 p. (B)*   *TC 147 / SC 2*
Water quality — Determination of the chemical oxygen demand

ISO 6332:1988   *Ed. 2   4 p. (B)*   *TC 147 / SC 2*
Water quality — Determination of iron — Spectrometric method
using 1,10-phenanthroline

ISO 6333:1986   *Ed. 1   4 p. (B)*   *TC 147 / SC 2*
Water quality — Determination of manganese — Formaldoxime
spectrometric method

ISO 6341:1996   *Ed. 3   9 p. (E)*   *TC 147 / SC 5*
Water quality — Determination of the inhibition of the mobility of
Daphnia magna Straus (Cladocera, Crustacea) — Acute toxicity
test

ISO 6439:1990   *Ed. 2   7 p. (D)*   *TC 147 / SC 2*
Water quality — Determination of phenol
index — 4-Aminoantipyrine spectrometric methods after
distillation

ISO/DIS 6468   *Ed. 1*   *TC 147 / SC 2*
Water quality — Determination of certain organochlorine
insecticides, polychlorinated biphenyls and
chlorobenzenes — Gas chromatographic method after liquid-liquid
extraction

ISO 6595:1982   *Ed. 1   5 p. (C)*   *TC 147 / SC 2*
Water quality — Determination of total arsenic — Silver
diethyldithiocarbamate spectrophotometric method

ISO 6703-1:1984   *Ed. 1   11 p. (F)*   *TC 147 / SC 2*
Water quality — Determination of cyanide —
Part 1: Determination of total cyanide

ISO 6703-2:1984   *Ed. 1   11 p. (F)*   *TC 147 / SC 2*
Water quality — Determination of cyanide —
Part 2: Determination of easily liberatable cyanide

ISO 6703-3:1984   *Ed. 1   6 p. (C)*   *TC 147 / SC 2*
Water quality — Determination of cyanide —
Part 3: Determination of cyanogen chloride

ISO 6703-4:1985   *Ed. 1*   *6 p. (C)*   *TC 147 / SC 2*
Water quality — Determination of cyanide —
Part 4: Determination of cyanide by diffusion at pH 6

ISO 6777:1984   *Ed. 1*   *5 p. (C)*   *TC 147 / SC 2*
Water quality — Determination of nitrite — Molecular absorption
spectrometric method

ISO 6778:1984   *Ed. 1*   *5 p. (L)*   *TC 147 / SC 2*
Water quality — Determination of ammonium — Potentiometric
method

ISO 6878-1:1986   *Ed. 1*   *11 p. (F)*   *TC 147 / SC 2*
Water quality — Determination of phosphorus —
Part 1: Ammonium molybdate spectrometric method

ISO 7027:1990   *Ed. 2*   *6 p. (C)*   *TC 147 / SC 2*
Water quality — Determination of turbidity

ISO 7150-1:1984   *Ed. 1*   *7 p. (M)*   *TC 147 / SC 2*
Water quality — Determination of ammonium —
Part 1: Manual spectrometric method

ISO 7150-2:1986   *Ed. 1*   *7 p. (D)*   *TC 147 / SC 2*
Water quality — Determination of ammonium —
Part 2: Automated spectrometric method

ISO 7346-1:1984   *Ed. 1*   *9 p. (E)*   *TC 147 / SC 5*
Water quality — Determination of the acute lethal toxicity of
substances to a freshwater fish (Brachydanio rerio,
Hamilton-Buchanan (Teleostei, Cyprinidae)) —
Part 1: Static method

ISO/DIS 7346-1   *Ed. 2*   *11 p.*   *TC 147 / SC 5*
Water quality — Determination of the acute lethal toxicity of
substances to a freshwater fish (Brachydanio rerio
Hamilton-Buchanan (Teleostei, Cyprinidae)) —
Part 1: Static method (Revision of ISO 7346-1:1984)

ISO 7346-2:1984   *Ed. 1*   *9 p. (E)*   *TC 147 / SC 5*
Water quality — Determination of the acute lethal toxicity of
substances to a freshwater fish (Brachydanio rerio
Hamilton-Buchanan (Teleostei, Cyprinidae)) —
Part 2: Semi-static method

ISO/DIS 7346-2   *Ed. 2*   *11 p.*   *TC 147 / SC 5*
Water quality — Determination of the acute lethal toxicity of
substances to a freshwater fish (Brachydanio rerio
Hamilton-Buchanan (Teleostei, Cyprinidae)) —
Part 2: Semi-static method (Revision of ISO 7346-2:1984)

ISO 7346-3:1984   *Ed. 1*   *10 p. (E)*   *TC 147 / SC 5*
Water quality — Determination of the acute lethal toxicity of
substances to a freshwater fish (Brachydanio rerio
Hamilton-Buchanan (Teleostei, Cyprinidae)) —
Part 3: Flow-through method

ISO/DIS 7346-3   *Ed. 2*   *11 p.*   *TC 147 / SC 5*
Water quality — Determination of the acute lethal toxicity of
substances to a freshwater fish (Brachydanio rerio
Hamilton-Buchanan (Teleostei, Cyprinidae)) —
Part 3: Flow-through method (Revision of ISO 7346-3:1984)

ISO 7393-1:1985   *Ed. 1*   *8 p. (D)*   *TC 147 / SC 2*
Water quality — Determination of free chlorine and total
chlorine —
Part 1: Titrimetric method using N,N-diethyl-1,4-phenylenediamine

ISO 7393-2:1985   *Ed. 1*   *8 p. (D)*   *TC 147 / SC 2*
Water quality — Determination of free chlorine and total
chlorine —
Part 2: Colorimetric method using
N,N-diethyl-1,4-phenylenediamine, for routine control purposes

ISO 7393-3:1990   *Ed. 2*   *7 p. (D)*   *TC 147 / SC 2*
Water quality — Determination of free chlorine and total
chlorine —
Part 3: Iodometric titration method for the determination of total
chlorine

ISO 7827:1994   *Ed. 2*   *7 p. (D)*   *TC 147 / SC 5*
Water quality — Evaluation in an aqueous medium of the
'ultimate' aerobic biodegradability of organic
compounds — Method by analysis of dissolved organic carbon
(DOC)

ISO 7828:1985   *Ed. 1*   *6 p. (C)*   *TC 147 / SC 5*
Water quality — Methods of biological sampling — Guidance on
handnet sampling of aquatic benthic macro-invertebrates

ISO 7875-1:1984   *Ed. 1*   *6 p. (C)*   *TC 147 / SC 2*
Water quality — Determination of surfactants —
Part 1: Determination of anionic surfactants by the methylene
blue spectrometric method

ISO/DIS 7875-1   *Ed. 1*   *TC 147 / SC 2*
Water quality — Determination of surfactants —
Part 1: Determination of anionic surfactants by measurement of
the methylene blue index (MBAS) (Revision of ISO 7875-1:1984)

ISO 7875-2:1984   *Ed. 1*   *6 p. (C)*   *TC 147 / SC 2*
Water quality — Determination of surfactants —
Part 2: Determination of non-ionic surfactants using Dragendorff
reagent

ISO 7887:1994   *Ed. 2*   *8 p. (D)*   *TC 147 / SC 2*
Water quality — Examination and determination of colour

ISO 7888:1985   *Ed. 1*   *6 p. (C)*   *TC 147 / SC 2*
Water quality — Determination of electrical conductivity

ISO 7890-1:1986   *Ed. 1*   *5 p. (C)*   *TC 147 / SC 2*
Water quality — Determination of nitrate —
Part 1: 2,6-Dimethylphenol spectrometric method

ISO 7890-2:1986   *Ed. 1*   *4 p. (B)*   *TC 147 / SC 2*
Water quality — Determination of nitrate —
Part 2: 4-Fluorophenol spectrometric method after distillation

ISO 7890-3:1988   *Ed. 1*   *4 p. (B)*   *TC 147 / SC 2*
Water quality — Determination of nitrate —
Part 3: Spectrometric method using sulfosalicylic acid

ISO 7980:1986   *Ed. 1*   *3 p. (B)*   *TC 147 / SC 2*
Water quality — Determination of calcium and
magnesium — Atomic absorption spectrometric method

ISO/DIS 7981-1   *Ed. 1*   *18 p.*   *TC 147 / SC 2*
Water quality — Determination of six specified polynuclear
hydrocarbons —
Part 1: Thin layer chromatographic method with fluorescence
detection

ISO/DIS 7981-2   *Ed. 1*   *20 p.*   *TC 147 / SC 2*
Water quality — Determination of six specified polynuclear
hydrocarbons —
Part 2: High performance liquid chromatographic method with
fluorescence detection

ISO 8165-1:1992   *Ed. 1*   *7 p. (D)*   *TC 147 / SC 2*
Water quality — Determination of selected monovalent
phenols —
Part 1: Gas chromatographic method after enrichment by
extraction

ISO/DIS 8165-2   *Ed. 1*   *10 p.*   *TC 147 / SC 2*
Water quality — Determination of selected monovalent
phenols —
Part 2: Gas chromatographic method after derivatization with
pentafluorobenzoyl chloride

ISO 8192:1986   *Ed. 1*   *9 p. (E)*   *TC 147 / SC 5*
Water quality — Test for inhibition of oxygen consumption by
activated sludge

ISO 8245:1987   *Ed. 1*   *5 p. (C)*   *TC 147 / SC 2*
Water quality — Guidelines for the determination of total organic
carbon (TOC)

ISO 8265:1988   *Ed. 1*   *9 p. (E)*   *TC 147 / SC 5*
Water quality — Design and use of quantitative samplers for
benthic macro-invertebrates on stony substrata in shallow
freshwaters

13.060.40

ISO 8288:1986   *Ed. 2   11 p. (F)   TC 147 / SC 2*
Water quality — Determination of cobalt, nickel, copper, zinc,
cadmium and lead — Flame atomic absorption spectrometric
methods

ISO 8466-1:1990   *Ed. 1   8 p. (D)   TC 147 / SC 7*
Water quality — Calibration and evaluation of analytical methods
and estimation of performance characteristics —
Part 1: Statistical evaluation of the linear calibration function

ISO 8466-2:1993   *Ed. 1   12 p. (F)   TC 147 / SC 7*
Water quality — Calibration and evaluation of analytical methods
and estimation of performance characteristics —
Part 2: Calibration strategy for non-linear second order calibration
functions

ISO 8467:1993   *Ed. 2   4 p. (B)   TC 147 / SC 2*
Water quality — Determination of permanganate index

ISO 8692:1989   *Ed. 1   6 p. (C)   TC 147 / SC 5*
Water quality — Fresh water algal growth inhibition test with
Scenedesmus subspicatus and Selenastrum capricornutum

ISO 9174:1990   *Ed. 1   6 p. (C)   TC 147 / SC 2*
Water quality — Determination of total chromium — Atomic
absorption spectrometric methods

ISO 9280:1990   *Ed. 2   5 p. (C)   TC 147 / SC 2*
Water quality — Determination of sulfate — Gravimetric method
using barium chloride

ISO 9297:1989   *Ed. 1   4 p. (B)   TC 147 / SC 2*
Water quality — Determination of chloride — Silver nitrate
titration with chromate indicator (Mohr's method)

ISO 9390:1990   *Ed. 1   4 p. (B)   TC 147 / SC 2*
Water quality — Determination of borate — Spectrometric
method using azomethine-H

ISO 9391:1993   *Ed. 1   13 p. (G)   TC 147 / SC 5*
Water quality — Sampling in deep waters for
macro-invertebrates — Guidance on the use of colonization,
qualitative and quantitative samplers

ISO 9408:1991   *Ed. 1   10 p. (E)   TC 147 / SC 5*
Water quality — Evaluation in an aqueous medium of the
'ultimate' aerobic biodegradability of organic
compounds — Method by determining the oxygen demand in a
closed respirometer

          Technical Corrigendum 1:1992 to ISO 9408:1991
                    Ed. 1   2 p.   TC 147 / SC 5

ISO 9439:1990   *Ed. 1   9 p. (E)   TC 147 / SC 5*
Water quality — Evaluation in an aqueous medium of the
'ultimate' aerobic biodegradability of organic
compounds — Method by analysis of released carbon dioxide

          Technical Corrigendum 1:1992 to ISO 9439:1990
                    Ed. 1   1 p.   TC 147 / SC 5

ISO 9509:1989   *Ed. 1   6 p. (C)   TC 147 / SC 5*
Water quality — Method for assessing the inhibition of
nitrification of activated sludge micro-organisms by chemicals and
waste waters

ISO 9562:1989   *Ed. 1   8 p. (D)   TC 147 / SC 2*
Water quality — Determination of adsorbable organic halogens
(AOX)

ISO 9696:1992   *Ed. 1   11 p. (F)   TC 147*
Water quality — Measurement of gross alpha activity in
non-saline water — Thick source method

ISO 9697:1992   *Ed. 1   10 p. (E)   TC 147*
Water quality — Measurement of gross beta activity in non-saline
water

ISO 9698:1989   *Ed. 1   7 p. (D)   TC 147*
Water quality — Determination of tritium activity
concentration — Liquid scintillation counting method

ISO 9887:1992   *Ed. 1   8 p. (D)   TC 147 / SC 5*
Water quality — Evaluation of the aerobic biodegradability of
organic compounds in an aqueous medium — Semi-continuous
activated sludge method (SCAS)

ISO 9888:1991   *Ed. 1   7 p. (D)   TC 147 / SC 5*
Water quality — Evaluation of the aerobic biodegradability of
organic compounds in an aqueous medium — Static test
(Zahn-Wellens method)

ISO 9963-1:1994   *Ed. 1   6 p. (C)   TC 147 / SC 2*
Water quality — Determination of alkalinity —
Part 1: Determination of total and composite alkalinity

ISO 9963-2:1994   *Ed. 1   7 p. (D)   TC 147 / SC 2*
Water quality — Determination of alkalinity —
Part 2: Determination of carbonate alkalinity

ISO 9964-1:1993   *Ed. 1   4 p. (B)   TC 147 / SC 2*
Water quality — Determination of sodium and potassium —
Part 1: Determination of sodium by atomic absorption
spectrometry

ISO 9964-2:1993   *Ed. 1   4 p. (B)   TC 147 / SC 2*
Water quality — Determination of sodium and potassium —
Part 2: Determination of potassium by atomic absorption
spectrometry

ISO 9964-3:1993   *Ed. 1   5 p. (C)   TC 147 / SC 2*
Water quality — Determination of sodium and potassium —
Part 3: Determination of sodium and potassium by flame emission
spectrometry

ISO 9965:1993   *Ed. 1   5 p. (C)   TC 147 / SC 2*
Water quality — Determination of selenium — Atomic
absorption spectrometric method (hydride technique)

ISO 10048:1991   *Ed. 1   5 p. (C)   TC 147 / SC 2*
Water quality — Determination of nitrogen — Catalytic digestion
after reduction with Devarda's alloy

ISO 10229:1994   *Ed. 1   12 p. (F)   TC 147 / SC 5*
Water quality — Determination of the prolonged toxicity of
substances to freshwater fish — Method for evaluating the
effects of substances on the growth rate of rainbow trout
(Oncorhynchus mykiss Walbaum (Teleostei, Salmonidae))

ISO 10253:1995   *Ed. 1   8 p. (D)   TC 147 / SC 5*
Water quality — Marine algal growth inhibition test with
Skeletonema costatum and Phaeodactylum tricornutum

ISO 10260:1992   *Ed. 1   6 p. (C)   TC 147 / SC 2*
Water quality — Measurement of biochemical
parameters — Spectrometric determination of the chlorophyll-a
concentration

ISO/DIS 10301   *Ed. 1   38 p.   TC 147 / SC 2*
Water quality — Determination of highly volatile halogenated
hydrocarbons — Gas-chromatographic methods

ISO 10304-1:1992   *Ed. 1   12 p. (F)   TC 147 / SC 2*
Water quality — Determination of dissolved fluoride, chloride,
nitrite, orthophosphate, bromide, nitrate and sulfate ions, using
liquid chromatography of ions —
Part 1: Method for water with low contamination

ISO 10304-2:1995   *Ed. 1   18 p. (J)   TC 147 / SC 2*
Water quality — Determination of dissolved anions by liquid
chromatography of ions —
Part 2: Determination of bromide, chloride, nitrate, nitrite,
orthophosphate and sulfate in waste water

ISO/DIS 10304-3   *Ed. 1   27 p.   TC 147 / SC 2*
Water quality — Determination of dissolved anions by liquid
chromatography of ions —
Part 3: Determination of chromate, iodide, sulfite, thiocyanate and
thiosulfate

ISO/DIS 10304-4   *Ed. 1   17 p.   TC 147 / SC 2*
Water quality — Determination of dissolved anions by liquid
chromatography of ions —
Part 4: Determination of dissolved chlorate, chloride and chlorite
in water with low contamination

ISO 10359-1:1992   *Ed. 1   6 p. (C)*   *TC 147 / SC 2*
Water quality — Determination of fluoride —
Part 1: Electrochemical probe method for potable and lightly
polluted water

ISO 10359-2:1994   *Ed. 1   7 p. (D)*   *TC 147 / SC 2*
Water quality — Determination of fluoride —
Part 2: Determination of inorganically bound total fluoride after
digestion and distillation

ISO 10523:1994   *Ed. 1   10 p. (E)*   *TC 147 / SC 2*
Water quality — Determination of pH

ISO 10530:1992   *Ed. 1   9 p. (E)*   *TC 147 / SC 2*
Water quality — Determination of dissolved
sulfide — Photometric method using methylene blue

ISO 10566:1994   *Ed. 1   5 p. (C)*   *TC 147 / SC 2*
Water quality — Determination of aluminium — Spectrometric
method using pyrocatechol violet

ISO 10634:1995   *Ed. 1   7 p. (D)*   *TC 147 / SC 5*
Water quality — Guidance for the preparation and treatment of
poorly water-soluble organic compounds for the subsequent
evaluation of their biodegradability in an aqueous medium

ISO/DIS 10703   *Ed. 1*   *TC 147*
Water quality — Determination of the activity concentration of
radionuclides by high resolution gamma-ray spectrometry

ISO 10707:1994   *Ed. 1   9 p. (E)*   *TC 147 / SC 5*
Water quality — Evaluation in an aqueous medium of the
'ultimate' aerobic biodegradability of organic
compounds — Method by analysis of biochemical oxygen
demand (closed bottle test)

ISO/DIS 10708   *Ed. 1   16 p.*   *TC 147 / SC 5*
Water quality — Evaluation in an aqueous medium of the
'ultimate' aerobic biodegradability of organic
compounds — Method by determining the biochemical oxygen
demand (two-phase closed bottle test)

ISO 10712:1995   *Ed. 1   9 p. (E)*   *TC 147 / SC 5*
Water quality — Pseudomonas putida growth inhibition test
(Pseudomonas cell multiplication inhibition test)

ISO 11083:1994   *Ed. 1   5 p. (C)*   *TC 147 / SC 2*
Water quality — Determination of chromium(VI) —
Spectrometric method using 1,5-diphenylcarbazide

ISO/DIS 11369   *Ed. 1   22 p.*   *TC 147 / SC 2*
Water quality — Determination of selected plant treatment
agents — Method using high performance liquid chromatography
with UV detection after solid-liquid extraction

ISO/DIS 11423-1   *Ed. 1*   *TC 147 / SC 2*
Water quality — Determination of benzene and some
derivatives —
Part 1: Head-space gas chromatographic method

ISO/DIS 11423-2   *Ed. 1*   *TC 147 / SC 2*
Water quality — Determination of benzene and some
derivatives —
Part 2: Gas chromatographic method after extraction

ISO/DIS 11732   *Ed. 1   19 p.*   *TC 147 / SC 2*
Water quality — Determination of ammonium nitrogen by flow
analysis and spectrometric detection

ISO 11733:1995   *Ed. 1   14 p. (G)*   *TC 147 / SC 5*
Water quality — Evaluation of the elimination and biodegradability
of organic compounds in an aqueous medium — Activated sludge
simulation test

ISO 11734:1995   *Ed. 1   13 p. (G)*   *TC 147 / SC 5*
Water quality — Evaluation of the 'ultimate' anaerobic
biodegradability of organic compounds in digested
sludge — Method by measurement of the biogas production

ISO/DIS 11885   *Ed. 1*   *TC 147 / SC 2*
Water quality — The determination of 33 elements by inductively
coupled plasma atomic emission spectroscopy

ISO/DIS 11905-1   *Ed. 1   13 p.*   *TC 147 / SC 2*
Water quality — Determination of nitrogen —
Part 1: Method using oxidative digestion with peroxodisulfate

ISO/DTR 11905-2   *Ed. 1*   *TC 147 / SC 2*
Water quality — Determination of nitrogen —
Part 2: Determination of bound nitrogen after oxidation and
combustion to nitrogen dioxide using chemiluminescent detection

ISO/DIS 11923   *Ed. 1*   *TC 147 / SC 2*
Water quality — Determination of suspended solids by filtration
through glass-fibre filters

ISO/DIS 11969   *Ed. 1   8 p.*   *TC 147 / SC 2*
Water quality — Determination of arsenic — Atomic absorption
spectrometric method (hydride technique)

ISO/DIS 12020   *Ed. 1*   *TC 147 / SC 2*
Water quality — Determination of aluminium — Atomic
absorption spectrometric method

ISO/DIS 13358   *Ed. 1   8 p.*   *TC 147 / SC 2*
Water quality — Determination of easily released sulfide

ISO/DIS 13395   *Ed. 1   18 p.*   *TC 147 / SC 2*
Water quality — Determination of nitrite nitrogen and nitrate
nitrogen and the sum of both by flow analysis (CFA and FIA) and
spectrometric detection

13.080   Soil quality. Pedology

ISO/DIS 10381-1   *Ed. 2   40 p.*   *TC 190 / SC 2*
Soil quality — Sampling —
Part 1: Guidance on the design of sampling programmes

ISO/DIS 10381-2   *Ed. 2   44 p.*   *TC 190 / SC 2*
Soil quality — Sampling —
Part 2: Guidance on sampling techniques

ISO/DIS 10381-3   *Ed. 2   44 p.*   *TC 190 / SC 2*
Soil quality — Sampling —
Part 3: Guidance on safety

ISO/DIS 10381-4   *Ed. 1   20 p.*   *TC 190 / SC 2*
Soil quality — Sampling —
Part 4: Guidance on the procedure for investigation of natural,
near natural and cultivated sites

ISO 10381-6:1993   *Ed. 1   4 p. (B)*   *TC 190 / SC 2*
Soil quality — Sampling —
Part 6: Guidance on the collection, handling and storage of soil for
the assessment of aerobic microbial processes in the laboratory

ISO 10390:1994   *Ed. 1   5 p. (C)*   *TC 190 / SC 3*
Soil quality — Determination of pH

ISO 10573:1995   *Ed. 1   13 p. (G)*   *TC 190 / SC 5*
Soil quality — Determination of water content in the unsaturated
zone — Neutron depth probe method

ISO 10693:1995   *Ed. 1   7 p. (D)*   *TC 190 / SC 3*
Soil quality — Determination of carbonate content — Volumetric
method

ISO 10694:1995   *Ed. 1   7 p. (D)*   *TC 190 / SC 3*
Soil quality — Determination of organic and total carbon after dry
combustion (elementary analysis)

ISO/TR 11046:1994   *Ed. 1   13 p. (G)*   *TC 190 / SC 3*
Soil quality — Determination of mineral oil content — Method by
infrared spectrometry and gas chromatographic method

ISO/DIS 11047   *Ed. 1   27 p.*   *TC 190 / SC 3*
Soil quality — Determination of cadmium, chromium, cobalt,
copper, lead, manganese, nickel and zinc — Flame and
electrothermal atomic absorption spectrometric methods

13.080

ISO 11048:1995   *Ed. 1   18 p. (J)*   *TC 190 / SC 3*
Soil quality — Determination of water-soluble and acid-soluble
sulfate

ISO/DIS 11074-1   *Ed. 1*   *TC 190 / SC 1*
Soil quality — Vocabulary —
Part 1: Terms and definitions relating to the protection and
pollution of the soil
Trilingual edition

ISO/DIS 11259   *Ed. 1   50 p.*   *TC 190 / SC 1*
Soil quality — Simplified soil description
Bilingual edition

ISO 11260:1994   *Ed. 1   10 p. (E)*   *TC 190 / SC 3*
Soil quality — Determination of effective cation exchange
capacity and base saturation level using barium chloride solution

      Draft Technical Corrigendum 1 to ISO 11260:1994
      Technical Corrigendum 1 to ISO 11260:1993
             Ed. 1   *TC 190 / SC 3*

ISO 11261:1995   *Ed. 1   4 p. (B)*   *TC 190 / SC 3*
Soil quality — Determination of total nitrogen — Modified
Kjeldahl method

ISO 11263:1994   *Ed. 1   5 p. (C)*   *TC 190 / SC 3*
Soil quality — Determination of phosphorus — Spectrometric
determination of phosphorus soluble in sodium hydrogen
carbonate solution

ISO 11265:1994   *Ed. 1   4 p. (B)*   *TC 190 / SC 3*
Soil quality — Determination of the specific electrical conductivity

      Draft Technical Corrigendum 1 to ISO 11265:1994
      Technical Corrigendum 1 to ISO 11265:1994
             Ed. 1   *TC 190 / SC 3*

ISO 11266:1994   *Ed. 1   6 p. (C)*   *TC 190 / SC 4*
Soil quality — Guidance on laboratory testing for biodegradation
of organic chemicals in soil under aerobic conditions

ISO 11268-1:1993   *Ed. 1   6 p. (C)*   *TC 190 / SC 4*
Soil quality — Effects of pollutants on earthworms (Eisenia
fetida) —
Part 1: Determination of acute toxicity using artificial soil
substrate

ISO/DIS 11268-2   *Ed. 1   15 p.*   *TC 190 / SC 4*
Soil quality — Effects of pollutants on earthworms (Eisenia
fetida) —
Part 2: Determination of effects on reproduction

ISO 11269-1:1993   *Ed. 1   9 p. (E)*   *TC 190 / SC 4*
Soil quality — Determination of the effects of pollutants on soil
flora —
Part 1: Method for the measurement of inhibition of root growth

ISO 11269-2:1995   *Ed. 1   7 p. (D)*   *TC 190 / SC 4*
Soil quality — Determination of the effects of pollutants on soil
flora —
Part 2: Effects of chemicals on the emergence and growth of
higher plants

ISO/DIS 11271   *Ed. 1   14 p.*   *TC 190 / SC 5*
Soil quality — Determination of redox potential — Field method

ISO/DIS 11272   *Ed. 1   9 p.*   *TC 190 / SC 5*
Soil quality — Determination of dry bulk density

ISO/DIS 11273-1   *Ed. 1   6 p.*   *TC 190 / SC 5*
Soil quality — Determination of aggregate strength —
Part 1: Tensile strength measurement (crushing test)

ISO/DIS 11274   *Ed. 1   30 p.*   *TC 190 / SC 5*
Soil quality — Determination of the water retention
characteristic — Laboratory methods

ISO/DIS 11275   *Ed. 1   9 p.*   *TC 190 / SC 5*
Soil quality — Determination of unsaturated hydraulic conductivity
and water retention characteristic — Wind's evaporation method

ISO 11276:1995   *Ed. 1   20 p. (K)*   *TC 190 / SC 5*
Soil quality — Determination of pore water
pressure — Tensiometer method

ISO/DIS 11277   *Ed. 1   45 p.*   *TC 190 / SC 5*
Soil quality — Determination of particle size distribution in mineral
soil material — Method by sieving and sedimentation following
removal of soluble salts, organic matter and carbonates

ISO/DIS 11461   *Ed. 1   6 p.*   *TC 190 / SC 5*
Soil quality — Determination of soil water content on a volume
basis — Gravimetric method

ISO 11464:1994   *Ed. 1   9 p. (E)*   *TC 190 / SC 3*
Soil quality — Pretreatment of samples for physico-chemical
analyses

ISO 11465:1993   *Ed. 1   3 p. (B)*   *TC 190 / SC 3*
Soil quality — Determination of dry matter and water content on
a mass basis — Gravimetric method

      Technical Corrigendum 1:1994 to ISO 11465:1993
             Ed. 1   1 p.   *TC 190 / SC 3*

ISO 11466:1995   *Ed. 1   6 p. (C)*   *TC 190 / SC 3*
Soil quality — Extraction of trace elements soluble in aqua regia

ISO/DIS 11508   *Ed. 1   4 p.*   *TC 190 / SC 5*
Soil quality — Determination of particle density

ISO 13536:1995   *Ed. 1   7 p. (D)*   *TC 190 / SC 3*
Soil quality — Determination of the potential cation exchange
capacity and exchangeable cations using barium chloride solution
buffered at pH = 8,1

ISO/DIS 13877   *Ed. 1   18 p.*   *TC 190 / SC 3*
Soil quality — Determination of polynuclear aromatic
hydrocarbons — Method using high-performance liquid
chromatography

ISO/DIS 13878   *Ed. 1   4 p.*   *TC 190 / SC 3*
Soil quality — Determination of total nitrogen content after dry
combustion ('elemental analysis')

ISO/DIS 14235   *Ed. 1   7 p.*   *TC 190 / SC 3*
Soil quality — Determination of organic carbon by sulfochromic
oxidation

ISO/DIS 14238   *Ed. 1*   *TC 190 / SC 4*
Soil quality — Biological methods — Determination of nitrogen
mineralization and nitrification in soils and the influence of
chemicals on these processes

ISO/DIS 14239   *Ed. 1*   *TC 190 / SC 4*
Soil quality — Laboratory incubation systems for measuring the
mineralization of organic chemicals in soil under aerobic conditions

ISO/DIS 14240-1   *Ed. 1*   *TC 190 / SC 4*
Soil quality — Determination of soil microbial biomass —
Part 1: Respiration method

ISO/DIS 14240-2   *Ed. 1*   *TC 190 / SC 4*
Soil quality — Determination of soil microbial biomass —
Part 2: Fumigation extraction method

ISO/DIS 14255   *Ed. 1   14 p.*   *TC 190 / SC 3*
Soil quality — Determination of soluble nitrogen fractions

ISO/DIS 14507   *Ed. 1   16 p.*   *TC 190 / SC 3*
Soil quality — Pretreatment of samples for the determination of
organic contaminants

67   FOOD TECHNOLOGY

67.020   Processes in the food industry

ISO 8086:1986   *Ed. 1*   *6 p. (C)*   *TC 34 / SC 5*
Dairy plant — Hygiene conditions — General guidance on
inspection and sampling procedures

67.040   Agricultural food products in general

ISO 1871:1975   *Ed. 1*   *5 p. (C)*   *TC 34*
Agricultural food products — General directions for the
determination of nitrogen by the Kjeldahl method

ISO 5498:1981   *Ed. 1*   *8 p. (D)*   *TC 34*
Agricultural food products — Determination of crude fibre
content — General method

ISO 6541:1981   *Ed. 1*   *4 p. (B)*   *TC 34*
Agricultural food products — Determination of crude fibre
content — Modified Scharrer method

ISO 7002:1986   *Ed. 1*   *17 p. (J)*   *TC 34*
Agricultural food products — Layout for a standard method of
sampling from a lot

ISO 11289:1993   *Ed. 1*   *4 p. (B)*   *TC 34 / SC 9*
Heat-processed foods in hermetically sealed
containers — Determination of pH

67.060   Cereals, pulses and derived products

ISO 520:1977   *Ed. 1*   *2 p. (A)*   *TC 34 / SC 4*
Cereals and pulses — Determination of the mass of 1000 grains

ISO 605:1991   *Ed. 2*   *5 p. (C)*   *TC 34 / SC 4*
Pulses — Determination of impurities, size, foreign odours,
insects, and species and variety — Test methods

ISO 711:1985   *Ed. 2*   *5 p. (C)*   *TC 34 / SC 4*
Cereals and cereal products — Determination of moisture content
(Basic reference method)

ISO 712:1985   *Ed. 2*   *3 p. (B)*   *TC 34 / SC 4*
Cereals and cereal products — Determination of moisture content
(Routine reference method)

ISO/DIS 712   *Ed. 3*   *7 p.*   *TC 34 / SC 4*
Cereals and cereal products — Determination of moisture
content — Routine reference method (Revision of ISO 712:1985)

ISO 950:1979   *Ed. 1*   *6 p. (C)*   *TC 34 / SC 4*
Cereals — Sampling (as grain)

ISO 951:1979   *Ed. 1*   *6 p. (C)*   *TC 34 / SC 4*
Pulses in bags — Sampling

ISO 2164:1975   *Ed. 1*   *4 p. (B)*   *TC 34 / SC 4*
Pulses — Determination of glycosidic hydrocyanic acid

ISO 2170:1980   *Ed. 2*   *8 p. (D)*   *TC 34 / SC 4*
Cereals and pulses — Sampling of milled products

ISO 2171:1993   *Ed. 3*   *5 p. (C)*   *TC 34 / SC 4*
Cereals and milled cereal products — Determination of total ash

ISO 3093:1982   *Ed. 2*   *5 p. (C)*   *TC 34 / SC 4*
Cereals — Determination of falling number

ISO 3983:1977   *Ed. 1*   *5 p. (C)*   *TC 34 / SC 4*
Cereals and cereal products — Determination of alpha-amylase
activity — Colorimetric method

ISO 4112:1990   *Ed. 2*   *3 p. (B)*   *TC 34 / SC 4*
Cereals and pulses — Guidance on measurement of the
temperature of grain stored in bulk

ISO 4174:1980   *Ed. 1*   *11 p. (F)*   *TC 34 / SC 4*
Cereals and pulses — Measurement of unit pressure losses due
to single-dimension air flow through a batch of grain

ISO/DIS 4174   *Ed. 1*   *8 p.*   *TC 34 / SC 4*
Cereals, oilseeds and pulses — Measurement of unit pressure
loss in one-dimensional air flow through bulk grain (Revision of
ISO 4174:1980)

ISO 5526:1986   *Ed. 1*   *17 p. (J)*   *TC 34 / SC 4*
Cereals, pulses and other food grains — Nomenclature
Trilingual edition

ISO 5527:1995   *Ed. 2*   *15 p. (H)*   *TC 34 / SC 4*
Cereals — Vocabulary
Bilingual edition

ISO 5529:1992   *Ed. 2*   *4 p. (B)*   *TC 34 / SC 4*
Wheat — Determination of sedimentation index — Zeleny test

ISO 5530-1:1988   *Ed. 1*   *8 p. (D)*   *TC 34 / SC 4*
Wheat flour — Physical characteristics of doughs —
Part 1: Determination of water absorption and rheological
properties using a farinograph

ISO/DIS 5530-1   *Ed. 1*   *15 p.*   *TC 34 / SC 4*
Wheat flour — Physical characteristics of doughs —
Part 1: Determination of water absorption and rheological
properties using a farinograph (Revision of ISO 5530-1:1988)

ISO 5530-2:1988   *Ed. 1*   *7 p. (D)*   *TC 34 / SC 4*
Wheat flour — Physical characteristics of doughs —
Part 2: Determination of rheological properties using an
extensograph

67.060

ISO/DIS 5530-2   *Ed. 1   15 p.*   *TC 34 / SC 4*
Wheat flour — Physical characteristics of doughs —
Part 2: Determination of rheological properties using an
extensograph (Revision of ISO 5530-2:1988)

ISO 5530-3:1988   *Ed. 1   6 p. (C)*   *TC 34 / SC 4*
Wheat flour — Physical characteristics of doughs —
Part 3: Determination of water absorption and rheological
properties using a valorigraph

ISO 5530-4:1991   *Ed. 2   12 p. (F)*   *TC 34 / SC 4*
Wheat flour — Physical characteristics of doughs —
Part 4: Determination of rheological properties using an
alveograph

    Technical Corrigendum 1:1992 to ISO 5530-4:1991
             Ed. 1   1 p.   *TC 34 / SC 4*

ISO 5531:1978   *Ed. 1   3 p. (B)*   *TC 34 / SC 4*
Wheat flour — Determination of wet gluten

ISO 6322-1:1981   *Ed. 1   8 p. (D)*   *TC 34 / SC 4*
Storage of cereals and pulses —
Part 1: General considerations in keeping cereals

ISO/DIS 6322-1   *Ed. 1*   *TC 34 / SC 4*
Storage of cereals and pulses —
Part 1: General considerations for the keeping of cereals
(Revision of ISO 6322-1:1982)

ISO 6322-2:1981   *Ed. 1   6 p. (C)*   *TC 34 / SC 4*
Storage of cereals and pulses —
Part 2: Essential requirements

ISO 6322-3:1989   *Ed. 2   6 p. (C)*   *TC 34 / SC 4*
Storage of cereals and pulses —
Part 3: Control of attack by pests

ISO 6540:1980   *Ed. 1   11 p. (F)*   *TC 34 / SC 4*
Maize — Determination of moisture content (on milled grains and
on whole grains)

ISO 6639-1:1986   *Ed. 1   2 p. (A)*   *TC 34 / SC 4*
Cereals and pulses — Determination of hidden insect
infestation —
Part 1: General principles

ISO 6639-2:1986   *Ed. 1   4 p. (B)*   *TC 34 / SC 4*
Cereals and pulses — Determination of hidden insect
infestation —
Part 2: Sampling

ISO 6639-3:1986   *Ed. 1   4 p. (B)*   *TC 34 / SC 4*
Cereals and pulses — Determination of hidden insect
infestation —
Part 3: Reference method

ISO 6639-4:1987   *Ed. 1   17 p. (J)*   *TC 34 / SC 4*
Cereals and pulses — Determination of hidden insect
infestation —
Part 4: Rapid methods

ISO 6644:1981   *Ed. 1   5 p. (C)*   *TC 34 / SC 4*
Cereals and milled cereal products — Automatic sampling by
mechanical means

ISO 6645:1981   *Ed. 3   2 p. (A)*   *TC 34 / SC 4*
Wheat flour — Determination of dry gluten

ISO 6646:1984   *Ed. 1   3 p. (B)*   *TC 34 / SC 4*
Rice — Determination of the yield of husked rice and milled rice

ISO 6647:1987   *Ed. 1   4 p. (B)*   *TC 34 / SC 4*
Rice — Determination of amylose content

ISO 6648:1993   *Ed. 2   4 p. (B)*   *TC 34 / SC 4*
Rice — Determination of viscoelastic properties at various stages
of cooking — Method using a viscoelastograph

ISO 6820:1985   *Ed. 1   4 p. (B)*   *TC 34 / SC 4*
Wheat flour and rye flour — General guidance on the drafting of
bread-making tests

ISO 7301:1988   *Ed. 1   12 p. (F)*   *TC 34 / SC 4*
Rice — Specification

ISO 7302:1982   *Ed. 1   3 p. (B)*   *TC 34 / SC 4*
Cereals and cereal products — Determination of total fat content

ISO 7304:1985   *Ed. 1   10 p. (E)*   *TC 34 / SC 4*
Durum wheat semolinas and alimentary pasta — Estimation of
cooking quality of spaghetti by sensory analysis

ISO 7305:1986   *Ed. 1   3 p. (B)*   *TC 34 / SC 4*
Milled cereal products — Determination of fat acidity

ISO/DIS 7305   *Ed. 2   7 p.*   *TC 34 / SC 4*
Milled cereal products — Determination of fat acidity (Revision of
ISO 7305:1986)

ISO 7495:1990   *Ed. 1   5 p. (C)*   *TC 34 / SC 4*
Wheat flour — Determination of wet gluten content by
mechanical means

ISO 7698:1990   *Ed. 1   6 p. (C)*   *TC 34 / SC 4*
Cereals, pulses and derived products — Enumeration of bacteria,
yeasts and moulds

ISO 7970:1989   *Ed. 1   9 p. (E)*   *TC 34 / SC 4*
Wheat — Specification

ISO 7971:1986   *Ed. 1   5 p. (C)*   *TC 34 / SC 4*
Cereals — Determination of bulk density, called 'mass per
hectolitre' (Reference method)

ISO 7971-2:1995   *Ed. 1   6 p. (C)*   *TC 34 / SC 4*
Cereals — Determination of bulk density, called 'mass per
hectolitre' —
Part 2: Routine method

ISO 7973:1992   *Ed. 1   8 p. (D)*   *TC 34 / SC 4*
Cereals and milled cereal products — Determination of the
viscosity of flour — Method using an amylograph

ISO 8981:1993   *Ed. 1   8 p. (D)*   *TC 34 / SC 4*
Wheat — Identification of varieties by electrophoresis

ISO 9648:1988   *Ed. 1   3 p. (B)*   *TC 34 / SC 4*
Sorghum — Determination of tannin content

ISO 11050:1993   *Ed. 1   11 p. (F)*   *TC 34 / SC 4*
Wheat flour and durum wheat semolina — Determination of
impurities of animal origin

ISO 11051:1994   *Ed. 1   12 p. (F)*   *TC 34 / SC 4*
Durum wheat (Triticum durum Desf.) — Specification

ISO 11052:1994   *Ed. 1   5 p. (C)*   *TC 34 / SC 4*
Durum wheat flour and semolina — Determination of yellow
pigment content

ISO/DIS 13690   *Ed. 1   20 p.*   *TC 34 / SC 4*
Cereals and pulses — Sampling (Revision of ISO 950:1979, ISO
951:1979 and ISO 2170:1980)

67.080   Fruits. Vegetables

ISO 750:1981   *Ed. 1   3 p. (B)*   *TC 34 / SC 3*
Fruit and vegetable products — Determination of titratable acidity

ISO 751:1981   *Ed. 1   2 p. (A)*   *TC 34 / SC 3*
Fruit and vegetable products — Determination of water-insoluble
solids content

ISO 762:1982   *Ed. 1   2 p. (A)*   *TC 34 / SC 3*
Fruit and vegetable products — Determination of mineral
impurities content

ISO 763:1982   *Ed. 1*   *2 p. (A)*   *TC 34 / SC 3*
Fruit and vegetable products — Determination of ash insoluble in hydrochloric acid

ISO 874:1980   *Ed. 1*   *3 p. (B)*   *TC 34 / SC 14*
Fresh fruits and vegetables — Sampling

ISO 1026:1982   *Ed. 1*   *4 p. (B)*   *TC 34 / SC 3*
Fruit and vegetable products — Determination of dry matter content by drying under reduced pressure and of water content by azeotropic distillation

ISO 1842:1991   *Ed. 2*   *2 p. (A)*   *TC 34 / SC 3*
Fruit and vegetable products — Determination of pH

ISO 1956-1:1982   *Ed. 1*   *22 p. (L)*   *TC 34 / SC 14*
Fruits and vegetables — Morphological and structural terminology —
Part 1
Trilingual edition

ISO 1956-2:1989   *Ed. 1*   *31 p. (P)*   *TC 34 / SC 14*
Fruits and vegetables — Morphological and structural terminology —
Part 2
Trilingual edition

ISO 2169:1981   *Ed. 2*   *5 p. (C)*   *TC 34 / SC 14*
Fruits and vegetables — Physical conditions in cold stores — Definitions and measurement

ISO 2173:1978   *Ed. 1*   *4 p. (B)*   *TC 34 / SC 3*
Fruit and vegetable products — Determination of soluble solids content — Refractometric method

ISO 2447:1974   *Ed. 1*   *2 p. (A)*   *TC 34 / SC 3*
Fruit and vegetable products — Determination of tin

ISO 2448:1973   *Ed. 1*   *3 p. (B)*   *TC 34 / SC 3*
Fruit and vegetable products — Determination of ethanol

ISO 3094:1974   *Ed. 1*   *2 p. (A)*   *TC 34 / SC 3*
Fruit and vegetable products — Determination of copper content — Photometric method

ISO 3659:1977   *Ed. 1*   *4 p. (B)*   *TC 34 / SC 14*
Fruits and vegetables — Ripening after cold storage

ISO 5515:1979   *Ed. 1*   *3 p. (B)*   *TC 34 / SC 3*
Fruits, vegetables and derived products — Decomposition of organic matter prior to analysis — Wet method

ISO 5516:1978   *Ed. 1*   *2 p. (A)*   *TC 34 / SC 3*
Fruits, vegetables and derived products — Decomposition of organic matter prior to analysis — Ashing method

ISO 5517:1978   *Ed. 1*   *3 p. (B)*   *TC 34 / SC 3*
Fruits, vegetables and derived products — Determination of iron content — 1,10- Phenanthroline photometric method

ISO 5518:1978   *Ed. 1*   *3 p. (B)*   *TC 34 / SC 3*
Fruits, vegetables and derived products — Determination of benzoic acid content — Spectrophotometric method

ISO 5519:1978   *Ed. 1*   *6 p. (C)*   *TC 34 / SC 3*
Fruits, vegetables and derived products — Determination of sorbic acid content

ISO 5520:1981   *Ed. 1*   *3 p. (B)*   *TC 34 / SC 3*
Fruits, vegetables and derived products — Determination of alkalinity of total ash and of water-soluble ash

ISO 5521:1981   *Ed. 1*   *2 p. (A)*   *TC 34 / SC 3*
Fruits, vegetables and derived products — Qualitative method for the detection of sulphur dioxide

ISO 5522:1981   *Ed. 1*   *7 p. (D)*   *TC 34 / SC 3*
Fruits, vegetables and derived products — Determination of total sulphur dioxide content

ISO 5523:1981   *Ed. 1*   *3 p. (B)*   *TC 34 / SC 3*
Liquid fruit and vegetable products — Determination of sulphur dioxide content (Routine method)

ISO 6557-1:1986   *Ed. 1*   *3 p. (B)*   *TC 34 / SC 3*
Fruits, vegetables and derived products — Determination of ascorbic acid —
Part 1: Reference method

ISO 6557-2:1984   *Ed. 1*   *4 p. (B)*   *TC 34 / SC 3*
Fruits, vegetables and derived products — Determination of ascorbic acid content —
Part 2: Routine methods

ISO 6558-2:1992   *Ed. 1*   *4 p. (B)*   *TC 34 / SC 3*
Fruits, vegetables and derived products — Determination of carotene content —
Part 2: Routine methods

ISO 6560:1983   *Ed. 1*   *3 p. (B)*   *TC 34 / SC 3*
Fruit and vegetable products — Determination of benzoic acid content (benzoic acid contents greater than 200 mg per litre or per kilogram) — Molecular absorption spectrometric method

ISO 6561:1983   *Ed. 1*   *3 p. (B)*   *TC 34 / SC 3*
Fruits, vegetables and derived products — Determination of cadmium content — Flameless atomic absorption spectrometric method

ISO/DIS 6562-2   *Ed. 1*   *TC 34 / SC 3*
Fruits, vegetables and derived products — Determination of pectic substances content —
Part 2: Spectrometric method with metahydroxydiphenyl reagent

ISO 6632:1981   *Ed. 1*   *9 p. (E)*   *TC 34 / SC 3*
Fruits, vegetables and derived products — Determination of volatile acidity

ISO 6633:1984   *Ed. 1*   *3 p. (B)*   *TC 34 / SC 3*
Fruits, vegetables and derived products — Determination of lead content — Flameless atomic absorption spectrometric method

ISO 6634:1982   *Ed. 1*   *5 p. (C)*   *TC 34 / SC 3*
Fruits, vegetables and derived products — Determination of arsenic content — Silver diethyldithiocarbamate spectrophotometric method

ISO 6635:1984   *Ed. 1*   *4 p. (B)*   *TC 34 / SC 3*
Fruits, vegetables and derived products — Determination of nitrite and nitrate content — Molecular absorption spectrometric method

ISO 6636-1:1986   *Ed. 1*   *3 p. (B)*   *TC 34 / SC 3*
Fruits, vegetables and derived products — Determination of zinc content —
Part 1: Polarographic method

ISO 6636-2:1981   *Ed. 1*   *4 p. (B)*   *TC 34 / SC 3*
Fruits, vegetables and derived products — Determination of zinc content —
Part 2: Atomic absorption spectrometric method

ISO 6636-3:1983   *Ed. 1*   *3 p. (B)*   *TC 34 / SC 3*
Fruit and vegetable products — Determination of zinc content —
Part 3: Dithizone spectrometric method

ISO 6637:1984   *Ed. 1*   *5 p. (C)*   *TC 34 / SC 3*
Fruits, vegetables and derived products — Determination of mercury content — Flameless atomic absorption method

ISO 6638-1:1985   *Ed. 1*   *3 p. (B)*   *TC 34 / SC 3*
Fruit and vegetable products — Determination of formic acid content —
Part 1: Gravimetric method

ISO 6638-2:1984   *Ed. 1*   *3 p. (B)*   *TC 34 / SC 3*
Fruit and vegetable products — Determination of formic acid content —
Part 2: Routine method

ISO 6661:1983   *Ed. 1*   *4 p. (B)*   *TC 34 / SC 14*
Fresh fruits and vegetables — Arrangement of parallelepipedic packages in land transport vehicles

67.080

ISO 6949:1988	*Ed. 1*	*6 p. (C)*	*TC 34 / SC 14*

Fruits and vegetables — Principles and techniques of the
controlled atmosphere method of storage

ISO 7466:1986   *Ed. 1*   *3 p. (B)*   *TC 34 / SC 3*
Fruit and vegetable products — Determination of
5-hydroxymethylfurfural (5-HMF) content

ISO 7558:1988   *Ed. 1*   *5 p. (C)*   *TC 34 / SC 14*
Guide to the prepacking of fruits and vegetables

ISO/DIS 7563   *Ed. 1*   *TC 34 / SC 14*
Fresh fruits and vegetables — Vocabulary
Bilingual edition

ISO 7952:1994   *Ed. 1*   *5 p. (C)*   *TC 34 / SC 3*
Fruits, vegetables and derived products — Determination of
copper content — Method using flame atomic absorption
spectrometry

ISO 8129-1:1984   *Ed. 1*   *2 p. (A)*   *TC 34 / SC 3*
Fruits, vegetables and derived products — Determination of
alcohol-insoluble solids content —
Part 1: Method for fresh or quick-frozen maize

ISO 8129-2:1984   *Ed. 1*   *2 p. (A)*   *TC 34 / SC 3*
Fruits, vegetables and derived products — Determination of
alcohol-insoluble solids content —
Part 2: Method for fresh or quick-frozen peas

ISO 9526:1990   *Ed. 1*   *3 p. (B)*   *TC 34 / SC 3*
Fruits, vegetables and derived products — Determination of iron
content by flame atomic absorption spectrometry

67.080.10 Fruits and derived products

ISO 873:1980   *Ed. 1*   *5 p. (C)*   *TC 34 / SC 14*
Peaches — Guide to cold storage

ISO 931:1980   *Ed. 1*   *3 p. (B)*   *TC 34 / SC 14*
Green bananas — Guide to storage and transport

ISO 1134:1993   *Ed. 2*   *8 p. (D)*   *TC 34 / SC 14*
Pears — Cold storage

ISO 1212:1995   *Ed. 2*   *8 p. (D)*   *TC 34 / SC 14*
Apples — Cold storage

ISO 1838:1993   *Ed. 2*   *4 p. (B)*   *TC 34 / SC 14*
Fresh pineapples — Storage and transport

ISO 1955:1982   *Ed. 1*   *3 p. (B)*   *TC 34 / SC 3*
Citrus fruits and derived products — Determination of essential
oils content (Reference method)

ISO 1990-1:1982   *Ed. 1*   *11 p. (F)*   *TC 34 / SC 14*
Fruits — Nomenclature — First list
Trilingual edition

ISO 1990-2:1985   *Ed. 1*   *5 p. (C)*   *TC 34 / SC 14*
Fruits — Nomenclature — Second list
Trilingual edition

ISO 2168:1974   *Ed. 1*   *5 p. (C)*   *TC 34 / SC 14*
Table grapes — Guide to cold storage

ISO 2295:1974   *Ed. 1*   *4 p. (B)*   *TC 34 / SC 14*
Avocados — Guide for storage and transport

ISO 2826:1974   *Ed. 1*   *3 p. (B)*   *TC 34 / SC 14*
Apricots — Guide to cold storage

ISO 3631:1978   *Ed. 1*   *9 p. (E)*   *TC 34 / SC 14*
Citrus fruits — Guide to storage

ISO 3959:1977   *Ed. 1*   *6 p. (C)*   *TC 34 / SC 14*
Green bananas — Ripening conditions

ISO 4125:1991   *Ed. 2*   *5 p. (C)*   *TC 34 / SC 13*
Dry fruits and dried fruits — Definitions and nomenclature
Trilingual edition

ISO 6477:1988   *Ed. 1*   *4 p. (B)*   *TC 34 / SC 13*
Cashew kernels — Specification

ISO 6478:1990   *Ed. 1*   *2 p. (A)*   *TC 34 / SC 13*
Peanuts — Specification

ISO 6479:1984   *Ed. 1*   *7 p. (D)*   *TC 34 / SC 13*
Shelled sweet kernels of apricots — Specification

ISO 6660:1993   *Ed. 2*   *3 p. (B)*   *TC 34 / SC 14*
Mangoes — Cold storage

ISO 6662:1983   *Ed. 1*   *4 p. (B)*   *TC 34 / SC 14*
Plums — Guide to cold storage

ISO 6664:1983   *Ed. 1*   *3 p. (B)*   *TC 34 / SC 14*
Bilberries and blueberries — Guide to cold storage

ISO 6665:1983   *Ed. 1*   *3 p. (B)*   *TC 34 / SC 14*
Strawberries — Guide to cold storage

ISO 6755:1984   *Ed. 1*   *4 p. (B)*   *TC 34 / SC 13*
Dried sour cherries — Specification

ISO 6756:1984   *Ed. 1*   *5 p. (C)*   *TC 34 / SC 13*
Decorticated stone pine nuts — Specification

ISO 6757:1984   *Ed. 1*   *4 p. (B)*   *TC 34 / SC 13*
Decorticated kernels of mahaleb cherries — Specification

ISO 7701:1994   *Ed. 2*   *10 p. (E)*   *TC 34 / SC 13*
Dried apples — Specification and test methods

ISO 7702:1995   *Ed. 2*   *10 p. (E)*   *TC 34 / SC 13*
Dried pears — Specification and test methods

ISO 7703:1995   *Ed. 2*   *10 p. (E)*   *TC 34 / SC 13*
Dried peaches — Specification and test methods

ISO 7907:1987   *Ed. 1*   *4 p. (B)*   *TC 34 / SC 13*
Carob — Specification

ISO 7908:1991   *Ed. 1*   *6 p. (C)*   *TC 34 / SC 13*
Dried sweet cherries — Specification

ISO 7910:1991   *Ed. 1*   *6 p. (C)*   *TC 34 / SC 13*
Dried mulberries — Specification

ISO 7911:1991   *Ed. 1*   *7 p. (D)*   *TC 34 / SC 13*
Unshelled pine nuts — Specification

ISO 7920:1984   *Ed. 1*   *2 p. (A)*   *TC 34 / SC 14*
Sweet cherries and sour cherries — Guide to cold storage and
refrigerated transport

ISO 8682:1987   *Ed. 1*   *6 p. (C)*   *TC 34 / SC 14*
Apples — Storage in controlled atmospheres

ISO 9833:1993   *Ed. 1*   *2 p. (A)*   *TC 34 / SC 14*
Melons — Cold storage and refrigerated transport

67.080.20 Vegetables and derived products

ISO 949:1987   *Ed. 2*   *2 p. (A)*   *TC 34 / SC 14*
Cauliflowers — Guide to cold storage and refrigerated transport

ISO 1673:1991   *Ed. 2*   *5 p. (C)*   *TC 34 / SC 14*
Onions — Guide to storage

ISO 1991-1:1982   *Ed. 1*   *9 p. (E)*   *TC 34 / SC 14*
Vegetables — Nomenclature — First list
Trilingual edition

ISO 1991-2:1995   *Ed. 2*   *5 p. (C)*   *TC 34 / SC 14*
Vegetables — Nomenclature —
Part 2: Second list
Bilingual edition

ISO 2165:1974   *Ed. 1*   *2 p. (A)*   *TC 34 / SC 14*
Ware potatoes — Guide to storage

ISO 2166:1981   *Ed. 2*   *2 p. (A)*   *TC 34 / SC 14*
Carrots — Guide to storage

ISO 2167:1991   *Ed. 3*   *3 p. (B)*   *TC 34 / SC 14*
Round-headed cabbage — Guide to cold storage and refrigerated
transport

ISO 3634:1979   *Ed. 1*   *3 p. (B)*   *TC 34 / SC 3*
Vegetable products — Determination of chloride content

ISO 4186:1980   *Ed. 1*   *1 p. (A)*   *TC 34 / SC 14*
Asparagus — Guide to storage

ISO 4187:1980   *Ed. 1*   *2 p. (A)*   *TC 34 / SC 14*
Horse-radish — Guide to storage

ISO 5524:1991   *Ed. 2*   *3 p. (B)*   *TC 34 / SC 14*
Tomatoes — Guide to cold storage and refrigerated transport

ISO 5525:1986   *Ed. 2*   *4 p. (B)*   *TC 34 / SC 14*
Potatoes — Storage in the open (in clamps)

ISO 6000:1981   *Ed. 1*   *5 p. (C)*   *TC 34 / SC 14*
Round-headed cabbage — Storage in the open

ISO 6659:1981   *Ed. 1*   *4 p. (B)*   *TC 34 / SC 14*
Sweet pepper — Guide to refrigerated storage and transport

ISO 6663:1995   *Ed. 2*   *2 p. (A)*   *TC 34 / SC 14*
Garlic — Cold storage

ISO 6822:1984   *Ed 1*   *4 p. (B)*   *TC 34 / SC 14*
Potatoes, root vegetables and round-headed cabbages — Guide
to storage in silos using forced ventilation

ISO 6882:1981   *Ed. 1*   *2 p. (A)*   *TC 34 / SC 14*
Asparagus — Guide to refrigerated transport

ISO 7560:1995   *Ed. 2*   *4 p. (B)*   *TC 34 / SC 14*
Cucumbers — Storage and refrigerated transport

ISO 7561:1984   *Ed. 1*   *2 p. (A)*   *TC 34 / SC 14*
Cultivated mushrooms — Guide to cold storage and refrigerated
transport

ISO 7562:1990   *Ed. 1*   *3 p. (B)*   *TC 34 / SC 14*
Potatoes — Guidelines for storage in artificially ventilated stores

ISO 7922:1985   *Ed. 1*   *2 p. (A)*   *TC 34 / SC 14*
Leeks — Guide to cold storage and refrigerated transport

ISO 8683:1988   *Ed. 1*   *2 p. (A)*   *TC 34 / SC 14*
Lettuce — Guide to precooling and refrigerated transport

ISO 9376:1988   *Ed. 1*   *2 p. (A)*   *TC 34 / SC 14*
Early potatoes — Guide to cooling and refrigerated transport

ISO 9719:1995   *Ed. 1*   *2 p. (A)*   *TC 34 / SC 14*
Root vegetables — Cold storage and refrigerated transport

ISO 9930:1993   *Ed. 1*   *2 p. (A)*   *TC 34 / SC 14*
Green beans — Storage and refrigerated transport

67.100   Milk and milk products

67.100.10   Milk. Milk products

ISO 707:1985   *Ed. 1*   *28 p. (N)*   *TC 34 / SC 5*
Milk and milk products — Methods of sampling

ISO/DIS 707   *Ed. 2*   *41 p.*   *TC 34 / SC 5*
Milk and milk products — Guidance on sampling (Revision of ISO
707:1985)

ISO 1211:1984   *Ed. 1*   *8 p. (D)*   *TC 34 / SC 5*
Milk — Determination of fat content — Gravimetric method
(Reference method)

ISO 1546:1981   *Ed. 1*   *7 p. (D)*   *TC 34 / SC 5*
Procedure for milk recording for cows

ISO 1736:1985   *Ed. 2*   *8 p. (D)*   *TC 34 / SC 5*
Dried milk, dried whey, dried buttermilk and dried butter
serum — Determination of fat content — Gravimetric method
(Reference method)

ISO 1737:1985   *Ed. 2*   *8 p. (D)*   *TC 34 / SC 5*
Evaporated milk and sweetened condensed milk —
Determination of fat content — Gravimetric method (Reference
method)

ISO 1740:1991   *Ed. 2*   *5 p. (C)*   *TC 34 / SC 5*
Milk fat products and butter — Determination of fat acidity
(Reference method)

ISO 2446:1976   *Ed. 1*   *7 p. (D)*   *TC 34 / SC 5*
Milk — Determination of fat content (Routine method)

ISO 2450:1985   *Ed. 2*   *8 p. (D)*   *TC 34 / SC 5*
Cream — Determination of fat content — Gravimetric method
(Reference method)

ISO 2911:1976   *Ed. 1*   *3 p. (B)*   *TC 34 / SC 5*
Sweetened condensed milk — Determination of sucrose
content — Polarimetric method

ISO 3356:1975   *Ed. 1*   *3 p. (B)*   *TC 34 / SC 5*
Milk and dried milk, buttermilk and buttermilk powder, whey and
whey powder — Determination of phosphatase activity
(Reference method)

ISO 3594:1976   *Ed. 1*   *5 p. (C)*   *TC 34 / SC 5*
Milk fat — Detection of vegetable fat by gas-liquid
chromatography of sterols (Reference method)

ISO 3595:1976   *Ed. 1*   *5 p. (C)*   *TC 34 / SC 5*
Milk fat — Detection of vegetable fat by the phytosteryl acetate
test

ISO 3728:1977   *Ed. 1*   *3 p. (B)*   *TC 34 / SC 5*
Ice cream and milk ice — Determination of total solids content
(Reference method)

ISO 3976:1977   *Ed. 1*   *3 p. (B)*   *TC 34 / SC 5*
Anhydrous milk fat — Determination of peroxide value (Reference
method)

ISO 5538:1987   *Ed. 1*   *14 p. (G)*   *TC 34 / SC 5*
Milk and milk products — Sampling — Inspection by attributes

ISO 5541-1:1986   *Ed. 1*   *7 p. (D)*   *TC 34 / SC 5*
Milk and milk products — Enumeration of coliforms —
Part 1: Colony count technique at 30 degrees C

ISO 5541-2:1986   *Ed. 1*   *8 p. (D)*   *TC 34 / SC 5*
Milk and milk products — Enumeration of coliforms —
Part 2: Most probable number technique at 30 degrees C

ISO 5542:1984   *Ed. 1*   *5 p. (C)*   *TC 34 / SC 5*
Milk — Determination of protein content — Amido black
dye-binding method (Routine method)

ISO 5543:1986   *Ed. 1*   *8 p. (D)*   *TC 34 / SC 5*
Caseins and caseinates — Determination of fat
content — Gravimetric method (Reference method)

ISO 5544:1978   *Ed. 1*   *2 p. (A)*   *TC 34 / SC 5*
Caseins — Determination of ' fixed ash ' (Reference method)

ISO 5545:1978   *Ed. 1*   *2 p. (A)*   *TC 34 / SC 5*
Rennet caseins and caseinates — Determination of ash
(Reference method)

67.100.10

ISO 5546:1979   *Ed. 1   2 p. (A)*   *TC 34 / SC 5*
Caseins and caseinates — Determination of pH (Reference method)

ISO 5547:1978   *Ed. 1   2 p. (A)*   *TC 34 / SC 5*
Caseins — Determination of free acidity (Reference method)

ISO 5548:1980   *Ed. 1   3 p. (B)*   *TC 34 / SC 5*
Caseins and caseinates — Determination of lactose content — Photometric method

ISO 5549:1978   *Ed. 1   3 p. (B)*   *TC 34 / SC 5*
Caseins and caseinates — Determination of protein content (Reference method)

ISO 5550:1978   *Ed. 1   2 p. (A)*   *TC 34 / SC 5*
Caseins and caseinates — Determination of water content (Reference method)

ISO 5738:1980   *Ed. 1   6 p. (C)*   *TC 34 / SC 5*
Milk and milk products — Determination of copper content — Photometric reference method

ISO 5739:1983   *Ed. 1   8 p. (D)*   *TC 34 / SC 5*
Caseins and caseinates — Determination of scorched particles content

ISO 5764:1987   *Ed. 1   6 p. (C)*   *TC 34 / SC 5*
Milk — Determination of freezing point — Thermistor cryoscope method

ISO/DIS 6090   *Ed. 1   4 p.*   *TC 34 / SC 5*
Milk and dried milk, buttermilk and buttermilk powder, whey and whey powder — Detection of phosphatase activity

ISO 6091:1980   *Ed. 1   2 p. (A)*   *TC 34 / SC 5*
Dried milk — Determination of titratable acidity (Reference method)

ISO 6092:1980   *Ed. 1   2 p. (A)*   *TC 34 / SC 5*
Dried milk — Determination of titratable acidity (Routine method)

ISO 6610:1992   *Ed. 1   5 p. (C)*   *TC 34 / SC 5*
Milk and milk products — Enumeration of colony-forming units of micro-organisms — Colony-count technique at 30 degrees C

ISO 6611:1992   *Ed. 1   6 p. (C)*   *TC 34 / SC 5*
Milk and milk products — Enumeration of colony-forming units of yeasts and/or moulds — Colony-count technique at 25 degrees C

ISO 6730:1992   *Ed. 1   6 p. (C)*   *TC 34 / SC 5*
Milk — Enumeration of colony-forming units of psychrotrophic micro-organisms — Colony-count technique at 6,5 degrees C

ISO 6731:1989   *Ed. 1   3 p. (B)*   *TC 34 / SC 5*
Milk, cream and evaporated milk — Determination of total solids content (Reference method)

ISO 6732:1985   *Ed. 1   6 p. (C)*   *TC 34 / SC 5*
Milk and milk products — Determination of iron content — Spectrometric method (Reference method)

ISO/DIS 6733   *Ed. 1*   *TC 34 / SC 5*
Canned evaporated milk, caseins and caseinates — Determination of lead content — Spectrometric method (reference method)

ISO 6734:1989   *Ed. 1   3 p. (B)*   *TC 34 / SC 5*
Sweetened condensed milk — Determination of total solids content (Reference method)

ISO 6735:1985   *Ed. 1   8 p. (D)*   *TC 34 / SC 5*
Dried milk — Assessment of heat class — Heat-number reference method

ISO 6736:1982   *Ed. 1   6 p. (C)*   *TC 34 / SC 5*
Dried milk — Determination of nitrate and nitrite contents — Method by cadmium reduction and photometry

ISO 6740:1985   *Ed. 1   6 p. (C)*   *TC 34 / SC 5*
Dried whey — Determination of nitrate and nitrite contents — Method by cadmium reduction and spectrometry

ISO 6785:1985   *Ed. 1   13 p. (G)*   *TC 34 / SC 5*
Milk and milk products — Detection of salmonella

ISO 7208:1984   *Ed. 1   7 p. (D)*   *TC 34 / SC 5*
Skimmed milk, whey and buttermilk — Determination of fat content — Gravimetric method (Reference method)

ISO 7328:1984   *Ed. 1   8 p. (D)*   *TC 34 / SC 5*
Milk-based edible ices and ice-mixes — Determination of fat content — Roese-Gottlieb gravimetric method (Reference method)

ISO/DIS 7889   *Ed. 1*   *TC 34 / SC 5*
Yoghurt — Enumeration of characteristic micro-organisms — Colony count technique at 37 degrees C

ISO 8069:1986   *Ed. 1   4 p. (B)*   *TC 34 / SC 5*
Dried milk — Determination of lactic acid and lactates content — Enzymatic method

ISO 8070:1987   *Ed. 1   3 p. (B)*   *TC 34 / SC 5*
Dried milk — Determination of sodium and potassium contents — Flame emission spectrometric method

ISO 8151:1987   *Ed. 1   3 p. (B)*   *TC 34 / SC 5*
Dried milk — Determination of nitrate content — Method by cadmium reduction and spectrometry (Screening method)

ISO 8156:1987   *Ed. 1   7 p. (D)*   *TC 34 / SC 5*
Dried milk and dried milk products — Determination of insolubility index

ISO 8195:1987   *Ed. 1   6 p. (C)*   *TC 34 / SC 5*
Caseins and caseinates — Determination of nitrate and nitrite contents — Method by cadmium reduction and spectrometry

ISO/DIS 8196   *Ed. 1   13 p.*   *TC 34 / SC 5*
Milk — Indirect methods of analysis — Definition and evaluation of the overall accuracy, and its application to calibration procedures and quality control in the dairy laboratory

ISO 8197:1988   *Ed. 1   5 p. (C)*   *TC 34 / SC 5*
Milk and milk products — Sampling — Inspection by variables

ISO 8261:1989   *Ed. 1   5 p. (C)*   *TC 34 / SC 5*
Milk and milk products — Preparation of test samples and dilutions for microbiological examination

ISO 8262-1:1987   *Ed. 1   5 p. (C)*   *TC 34 / SC 5*
Milk products and milk-based foods — Determination of fat content by the Weibull- Berntrop gravimetric method (Reference method) —
Part 1: Infant foods

ISO 8262-2:1987   *Ed. 1   6 p. (C)*   *TC 34 / SC 5*
Milk products and milk-based foods — Determination of fat content by the Weibull- Berntrop gravimetric method (Reference method) —
Part 2: Edible ices and ice-mixes

ISO 8262-3:1987   *Ed. 1   5 p. (C)*   *TC 34 / SC 5*
Milk products and milk-based foods — Determination of fat content by the Weibull- Berntrop gravimetric method (Reference method) —
Part 3: Special cases

ISO 8381:1987   *Ed. 1   9 p. (E)*   *TC 34 / SC 5*
Milk-based infant foods — Determination of fat content — Röse-Gottlieb gravimetric method (Reference method)

ISO/DIS 8552   *Ed. 1   4 p.*   *TC 34 / SC 5*
Milk — Enumeration of psychrotrophic micro-organisms — Colony count technique at 21 degrees C (Rapid method)

ISO/DIS 8553   *Ed. 1   7 p.*   *TC 34 / SC 5*
Milk — Enumeration of micro-organisms — Plate loop technique at 30 degrees C

ISO 8967:1992   *Ed. 1   4 p. (B)*   *TC 34 / SC 5*
Dried milk and dried milk products — Determination of bulk density

ISO 9874:1992   *Ed. 1*   *5 p. (C)*   *TC 34 / SC 5*
Milk — Determination of total phosphorus content — Method
using molecular absorption spectrometry

ISO 10560:1993   *Ed. 1*   *12 p. (F)*   *TC 34 / SC 5*
Milk and milk products — Detection of Listeria monocytogenes

Technical Corrigendum 1:1994 to ISO 10560:1993
              Ed. 1   *1 p.*   *TC 34 / SC 5*

ISO/DIS 11813   *Ed. 1*   *7 p.*   *TC 34 / SC 5*
Milk and milk products — Determination of zinc
content — Method using atomic absorption spectrometry

ISO/DIS 11816-1   *Ed. 1*   *TC 34 / SC 5*
Milk and milk products — Determination of alkaline phosphatase
activity — Fluorimetric method —
Part 1: Method for milk and milk-based drinks

ISO 11865:1995   *Ed. 1*   *5 p. (C)*   *TC 34 / SC 5*
Instant whole milk powder — Determination of white flecks
number

ISO/DIS 11866-1   *Ed. 1*   *11 p.*   *TC 34 / SC 5*
Milk and milk products — Enumeration of presumptive
Escherichia coli —
Part 1: Most probable number technique

ISO/DIS 11866-2   *Ed. 1*   *14 p.*   *TC 34 / SC 5*
Milk and milk products — Enumeration of presumptive
Escherichia coli —
Part 2: Most probable number technique using
4-methylumbelliferyl-beta-D-glucuronide (MUG)

ISO/DIS 11866-3   *Ed. 1*   *11 p.*   *TC 34 / SC 5*
Milk and milk products — Enumeration of presumptive
Escherichia coli —
Part 3: Colony-count technique at 44 degrees C using
membranes (Supersedes ISO/DIS 9621)

ISO/DIS 11868   *Ed. 1*   *9 p.*   *TC 34 / SC 5*
Heat-treated milk — Determination of lactulose
content — Method using high-performance liquid chromatography

ISO/DIS 11869   *Ed. 1*   *TC 34 / SC 5*
Yoghurt — Determination of titratable acidity — Potentiometric
method

ISO/DIS 12081   *Ed. 1*   *5 p.*   *TC 34 / SC 5*
Milk — Determination of calcium content — Titrimetric method

ISO/DIS 13366-1   *Ed. 1*   *5 p.*   *TC 34 / SC 5*
Milk — Enumeration of somatic cells —
Part 1: Microscopic method (Reference method)

ISO/DIS 13366-2   *Ed. 1*   *15 p.*   *TC 34 / SC 5*
Milk — Enumeration of somatic cells —
Part 2: Method using an electronic particle counter

ISO/DIS 13366-3   *Ed. 1*   *6 p.*   *TC 34 / SC 5*
Milk — Enumeration of somatic cells —
Part 3: Fluoro-opto-electronic method

67.100.20   Butter. Cheese

ISO 1735:1987   *Ed. 2*   *8 p. (D)*   *TC 34 / SC 5*
Cheese and processed cheese products — Determination of fat
content — Gravimetric method (Reference method)

ISO 1738:1980   *Ed. 1*   *2 p. (A)*   *TC 34 / SC 5*
Butter — Determination of salt content (Reference method)

ISO 1739:1975   *Ed. 1*   *1 p. (A)*   *TC 34 / SC 5*
Butter — Determination of the refractive index of the fat
(Reference method)

ISO 1740:1991   *Ed. 2*   *5 p. (C)*   *TC 34 / SC 5*
Milk fat products and butter — Determination of fat acidity
(Reference method)

ISO 1854:1987   *Ed. 2*   *8 p. (D)*   *TC 34 / SC 5*
Whey cheese — Determination of fat content — Gravimetric
method (Reference method)

ISO 2920:1974   *Ed. 1*   *2 p. (A)*   *TC 34 / SC 5*
Whey cheese — Determination of dry matter content (Reference
method)

ISO 2962:1984   *Ed. 2*   *3 p. (B)*   *TC 34 / SC 5*
Cheese and processed cheese products — Determination of
total phosphorus content — Molecular absorption spectrometric
method

ISO 2963:1974   *Ed. 1*   *2 p. (A)*   *TC 34 / SC 5*
Cheese and processed cheese products — Determination of
citric acid content (Reference method)

ISO/DIS 2963   *Ed. 2*   *TC 34 / SC 5*
Cheese and processed cheese products — Determination of
citric acid content — Enzymatic method (Revision of ISO
2963:1974)

ISO 3433:1975   *Ed. 1*   *4 p. (B)*   *TC 34 / SC 5*
Cheese — Determination of fat content — Van Gulik method

ISO 3727:1977   *Ed. 1*   *3 p. (B)*   *TC 34 / SC 5*
Butter — Determination of water, solids-not-fat and fat contents
on the same test portion (Reference method)

ISO 4099:1984   *Ed. 2*   *6 p. (C)*   *TC 34 / SC 5*
Cheese — Determination of nitrate and nitrite
contents — Method by cadmium reduction and photometry

ISO 5534:1985   *Ed. 2*   *2 p. (A)*   *TC 34 / SC 5*
Cheese and processed cheese — Determination of total solids
content (Reference method)

ISO 5943:1988   *Ed. 2*   *2 p. (A)*   *TC 34 / SC 5*
Cheese and processed cheese products — Determination of
chloride content — Potentiometric titration method

ISO 6739:1988   *Ed. 2*   *6 p. (C)*   *TC 34 / SC 5*
Whey cheese — Determination of nitrate and nitrite
contents — Method by cadmium reduction and spectrometry

ISO 7238:1983   *Ed. 1*   *2 p. (A)*   *TC 34 / SC 5*
Butter — Determination of pH of the serum — Potentiometric
method

ISO 7586:1985   *Ed. 1*   *3 p. (B)*   *TC 34 / SC 5*
Butter — Determination of water dispersion value

ISO 9233:1991   *Ed. 1*   *10 p. (E)*   *TC 34 / SC 5*
Cheese and cheese rind — Determination of natamycin
content — Method by molecular absorption spectrometry and by
high-performance liquid chromatography

ISO/DIS 12082   *Ed. 1*   *TC 34 / SC 5*
Processed cheese and cheese products — Calculation of the
content of added citrate emulsifying agents and
acidifiers/pH-controlling agents, expressed as citric acid

67.120   Meat, meat products and other animal produce

67.120.10   Meat and meat products

ISO 936:1978   *Ed. 1*   *2 p. (A)*   *TC 34 / SC 6*
Meat and meat products — Determination of ash (Reference
method)

ISO/DIS 936   *Ed. 1*   *5 p.*   *TC 34 / SC 6*
Meat and meat products — Determination of total ash (Revision
of ISO 936:1978)

67.120.10

ISO 937:1978   *Ed. 1   3 p. (B)*   *TC 34 / SC 6*
Meat and meat products — Determination of nitrogen content
(Reference method)

ISO 1442:1973   *Ed. 1   2 p. (A)*   *TC 34 / SC 6*
Meat and meat products — Determination of moisture content

ISO/DIS 1442   *Ed. 2   3 p.*   *TC 34 / SC 6*
Meat and meat products — Determination of moisture content
(Reference method) (Revision of ISO 1442:1973)

ISO 1443:1973   *Ed. 1   2 p. (A)*   *TC 34 / SC 6*
Meat and meat products — Determination of total fat content

ISO/DIS 1443   *Ed. 2   6 p.*   *TC 34 / SC 6*
Meat and meat products — Determination of total fat content
(Revision of ISO 1443:1973)

ISO 1444:1973   *Ed. 1   2 p. (A)*   *TC 34 / SC 6*
Meat and meat products — Determination of free fat content

ISO/DIS 1444   *Ed. 2   4 p.*   *TC 34 / SC 6*
Meat and meat products — Determination of free fat content
(Revision of ISO 1444:1973)

ISO 1841:1981   *Ed. 1   3 p. (B)*   *TC 34 / SC 6*
Meat and meat products — Determination of chloride content
(Reference method)

ISO/DIS 1841-1   *Ed. 1*   *TC 34 / SC 6*
Meat and meat products — Determination of chloride
content —
Part 1: Volhard method (Revision of ISO 1841:1981)

ISO/DIS 1841-2   *Ed. 1*   *TC 34 / SC 6*
Meat and meat products — Determination of chloride
content —
Part 2: Potentiometric method

ISO 2293:1988   *Ed. 2   4 p. (B)*   *TC 34 / SC 6*
Meat and meat products — Enumeration of
micro-organisms — Colony count technique at 30 degrees C
(Reference method)

ISO 2294:1974   *Ed. 1   3 p. (B)*   *TC 34 / SC 6*
Meat and meat products — Determination of total phosphorus
content (Reference method)

ISO 2917:1974   *Ed. 1   3 p. (B)*   *TC 34 / SC 6*
Meat and meat products — Measurement of pH (Reference
method)

ISO 2918:1975   *Ed. 1   3 p. (B)*   *TC 34 / SC 6*
Meat and meat products — Determination of nitrite content
(Reference method)

ISO 3091:1975   *Ed. 1   5 p. (C)*   *TC 34 / SC 6*
Meat and meat products — Determination of nitrate content
(Reference method)

ISO 3100-1:1991   *Ed. 2   4 p. (B)*   *TC 34 / SC 6*
Meat and meat products — Sampling and preparation of test
samples —
Part 1: Sampling

ISO 3100-2:1988   *Ed. 1   4 p. (B)*   *TC 34 / SC 6*
Meat and meat products — Sampling and preparation of test
samples —
Part 2: Preparation of test samples for microbiological
examination

ISO 3496:1994   *Ed. 2   5 p. (C)*   *TC 34 / SC 6*
Meat and meat products — Determination of hydroxyproline
content

ISO 3565:1975   *Ed. 1   11 p. (F)*   *TC 34 / SC 6*
Meat and meat products — Detection of salmonellae (Reference
method)

ISO/DIS 3565   *Ed. 2*   *TC 34 / SC 6*
Meat and meat products — Methods for the detection of
Salmonella (Revision of ISO 3565:1975)

ISO 3811:1979   *Ed. 1   5 p. (C)*   *TC 34 / SC 6*
Meat and meat products — Detection and enumeration of
presumptive coliform bacteria and presumptive Escherichia
coli — (Reference method)

ISO/DIS 3811   *Ed. 1   11 p.*   *TC 34 / SC 6*
Meat and meat products — Detection and enumeration of
presumptive coliform bacteria (Reference method) (Revision, in
part, of ISO 3811:1979)

ISO 4133:1979   *Ed. 1   4 p. (B)*   *TC 34 / SC 6*
Meat and meat products — Determination of
glucono-delta-lactone content (Reference method)

ISO 4134:1978   *Ed. 1   4 p. (B)*   *TC 34 / SC 6*
Meat and meat products — Determination of L-(+)- glutamic acid
content — Reference method

ISO/DIS 4134   *Ed. 1   14 p.*   *TC 34 / SC 6*
Meat and meat products — Determination of L-(+)- glutamic acid
content — Reference method (Revision of ISO 4134:1978)

ISO 5552:1979   *Ed. 1   6 p. (C)*   *TC 34 / SC 6*
Meat and meat products — Detection and enumeration of
Enterobacteriaceae (Reference methods)

ISO 5553:1980   *Ed. 1   3 p. (B)*   *TC 34 / SC 6*
Meat and meat products — Detection of polyphosphates

ISO 5554:1978   *Ed. 1   5 p. (C)*   *TC 34 / SC 6*
Meat products — Determination of starch content (Reference
method)

ISO 6391:1988   *Ed. 1   5 p. (C)*   *TC 34 / SC 6*
Meat and meat products — Enumeration of Escherichia
coli — Colony count technique at 44 degrees C using membranes

ISO/DIS 6391   *Ed. 1   9 p.*   *TC 34 / SC 6*
Meat and meat products — Enumeration of Escherichia
coli — Colony-count technique at 44 degrees C using membranes
(Revision of ISO 6391:1988)

ISO/DIS 12074   *Ed. 1   7 p.*   *TC 34 / SC 6*
Meat and meat products — Enumeration of presumptive
Escherichia coli — Most probable number technique (Revision, in
part, of ISO 3811:1979)

ISO/DIS 13493   *Ed. 1   8 p.*   *TC 34 / SC 6*
Meat and meat products — Determination of chloramphenicol
content using liquid chromatography

ISO/DIS 13496   *Ed. 1   11 p.*   *TC 34 / SC 6*
Meat and meat products — Detection of colouring
agents — Method using thin-layer chromatography

ISO 13681:1995   *Ed. 1   6 p. (C)*   *TC 34 / SC 6*
Meat and meat products — Enumeration of yeasts and
moulds — Colony-count technique

ISO 13720:1995   *Ed. 1   7 p. (D)*   *TC 34 / SC 6*
Meat and meat products — Enumeration of Pseudomonas spp.

ISO 13721:1995   *Ed. 1   5 p. (C)*   *TC 34 / SC 6*
Meat and meat products — Enumeration of lactic acid
bacteria — Colony-count technique at 30 degrees C

          Draft Technical Corrigendum 1 to ISO 13721:1995
          Technical Corrigendum 1
                              Ed. 1   *TC 34 / SC 6*

ISO/DIS 13722   *Ed. 1*   *TC 34 / SC 6*
Meat and meat products - Enumeration of Brochothrix
thermosphacta — Colony-count technique between 22 degrees C
and 25 degrees C

ISO/DIS 13730   *Ed. 1   6 p.*   *TC 34 / SC 6*
Meat and meat products — Determination of total phosphorus
content — Spectrometric method

67.140   Tea. Coffee. Cocoa

67.140.10 Tea

ISO 1572:1980   *Ed. 2   2 p. (A)*   *TC 34 / SC 8*
Tea — Preparation of ground sample of known dry matter
content

ISO 1573:1980   *Ed. 2   2 p. (A)*   *TC 34 / SC 8*
Tea — Determination of loss in mass at 103 degrees C

ISO 1575:1987   *Ed. 3   2 p. (A)*   *TC 34 / SC 8*
Tea — Determination of total ash

ISO 1576:1988   *Ed. 2   2 p. (A)*   *TC 34 / SC 8*
Tea — Determination of water-soluble ash and water-insoluble
ash

ISO 1577:1987   *Ed. 2   2 p. (A)*   *TC 34 / SC 8*
Tea — Determination of acid-insoluble ash

ISO 1578:1975   *Ed. 1   2 p. (A)*   *TC 34 / SC 8*
Tea — Determination of alkalinity of water-soluble ash

ISO 1839:1980   *Ed. 1   3 p. (B)*   *TC 34 / SC 8*
Tea — Sampling

ISO 3103:1980   *Ed. 1   4 p. (B)*   *TC 34 / SC 8*
Tea — Preparation of liquor for use in sensory tests

ISO 3720:1986   *Ed. 3   2 p. (A)*   *TC 34 / SC 8*
Black tea — Definition and basic requirements

Technical Corrigendum 1:1992 to ISO 3720:1986
                          Ed. 1   1 p.   *TC 34 / SC 8*

ISO 6078:1982   *Ed. 1   22 p. (L)*   *TC 34 / SC 8*
Black tea — Vocabulary
Bilingual edition

ISO 6079:1990   *Ed. 1   2 p. (A)*   *TC 34 / SC 8*
Instant tea in solid form — Specification

ISO 6770:1982   *Ed. 1   6 p. (C)*   *TC 34 / SC 8*
Instant tea — Determination of free-flow and compacted bulk
densities

ISO 7513:1990   *Ed. 1   2 p. (A)*   *TC 34 / SC 8*
Instant tea in solid form — Determination of moisture content
(loss in mass at 103 degrees C)

ISO 7514:1990   *Ed. 1   2 p. (A)*   *TC 34 / SC 8*
Instant tea in solid form — Determination of total ash

ISO 7516:1984   *Ed. 1   4 p. (B)*   *TC 34 / SC 8*
Instant tea in solid form — Sampling

ISO 9768:1994   *Ed. 2   3 p. (B)*   *TC 34 / SC 8*
Tea — Determination of water extract

ISO 9884-1:1994   *Ed. 1   6 p. (C)*   *TC 34 / SC 8*
Tea sacks — Specification —
Part 1: Reference sack for palletized and containerized transport
of tea

ISO/DIS 9884-2   *Ed. 1   7 p.*   *TC 34 / SC 8*
Tea sacks — Specification —
Part 2: Performance specification for sacks for palletized and
containerized transport of tea

ISO 10727:1995   *Ed. 1   6 p. (C)*   *TC 34 / SC 8*
Tea and instant tea in solid form — Determination of caffeine
content — Method using high-performance liquid chromatography

ISO/DIS 11286   *Ed. 1   4 p.*   *TC 34 / SC 8*
Tea — Method for the classification of grades by particle size
analysis

67.140.20 Coffee and coffee substitutes

ISO 1446:1978   *Ed. 1   5 p. (C)*   *TC 34 / SC 15*
Green coffee — Determination of moisture content (Basic
reference method)

ISO 1447:1978   *Ed. 1   2 p. (A)*   *TC 34 / SC 15*
Green coffee — Determination of moisture content (Routine
method)

ISO 3509:1989   *Ed. 3   12 p. (F)*   *TC 34 / SC 15*
Coffee and its products — Vocabulary
Bilingual edition

ISO 3726:1983   *Ed. 1   2 p. (A)*   *TC 34 / SC 15*
Instant coffee — Determination of loss in mass at 70 degrees C
under reduced pressure

ISO 4052:1983   *Ed. 1   6 p. (C)*   *TC 34 / SC 15*
Coffee — Determination of caffeine content (Reference method)

ISO 4072:1982   *Ed. 1   3 p. (B)*   *TC 34 / SC 15*
Green coffee in bags — Sampling

ISO 4149:1980   *Ed. 1   2 p. (A)*   *TC 34 / SC 15*
Green coffee — Olfactory and visual examination and
determination of foreign matter and defects

ISO 4150:1991   *Ed. 2   6 p. (C)*   *TC 34 / SC 15*
Green coffee — Size analysis — Manual sieving

ISO 6667:1985   *Ed. 1   12 p. (F)*   *TC 34 / SC 15*
Green coffee — Determination of proportion of insect-damaged
beans

ISO 6668:1991   *Ed. 1   3 p. (B)*   *TC 34 / SC 15*
Green coffee — Preparation of samples for use in sensory
analysis

ISO 6669:1995   *Ed. 1   3 p. (B)*   *TC 34 / SC 15*
Green and roasted coffee — Determination of free-flow bulk
density of whole beans (Routine method)

ISO 6670:1983   *Ed. 1   6 p. (C)*   *TC 34 / SC 15*
Instant coffee in cases with liners — Sampling

ISO 6673:1983   *Ed. 1   2 p. (A)*   *TC 34 / SC 15*
Green coffee — Determination of loss in mass at 105 degrees C

ISO 7532:1985   *Ed. 1   2 p. (A)*   *TC 34 / SC 15*
Instant coffee — Size analysis

ISO 7534:1985   *Ed. 1   3 p. (B)*   *TC 34 / SC 15*
Instant coffee — Determination of insoluble matter content

ISO 8455:1986   *Ed. 1   3 p. (B)*   *TC 34 / SC 15*
Green coffee in bags — Guide to storage and transport

ISO 8460:1987   *Ed. 1   6 p. (C)*   *TC 34 / SC 15*
Instant coffee — Determination of free-flow and compacted bulk
densities

ISO 9116:1992   *Ed. 1   3 p. (B)*   *TC 34 / SC 15*
Green coffee — Guidance on methods of specification

ISO 10095:1992   *Ed. 1   6 p. (D)*   *TC 34 / SC 15*
Coffee — Determination of caffeine content — Method using
high-performance liquid chromatography

ISO 10470:1993   *Ed. 1   22 p. (L)*   *TC 34 / SC 15*
Green coffee — Defect reference chart

ISO 11292:1995   *Ed. 1   15 p. (H)*   *TC 34 / SC 15*
Instant coffee — Determination of free and total carbohydrate
contents — Method using high-performance anion-exchange
chromatography

ISO 11294:1994   *Ed. 1   3 p. (B)*   *TC 34 / SC 15*
Roasted ground coffee — Determination of moisture
content — Method by determination of loss in mass at 103
degrees C (Routine method)

67.140.20

ISO 11817:1994   *Ed. 1   5 p. (C)   TC 34 / SC 15*
Roasted ground coffee — Determination of moisture
content — Karl Fischer method (Reference method)

67.140.30 Cocoa. Chocolate

ISO 1114:1977   *Ed. 1   1 p. (A)   TC 34*
Cocoa beans — Cut test

ISO 2291:1980   *Ed. 2   2 p. (A)   TC 34*
Cocoa beans — Determination of moisture content (Routine
method)

ISO 2292:1973   *Ed. 1   5 p. (C)   TC 34*
Cocoa beans — Sampling

ISO 2451:1973   *Ed. 1   3 p. (B)   TC 34*
Cocoa beans — Specification

67.160   Beverages

67.160.20 Non-alcoholic beverages

ISO 2172:1983   *Ed. 1   3 p. (B)   TC 34 / SC 3*
Fruit juice — Determination of soluble solids
content — Pycnometric method

ISO 8128-1:1993   *Ed. 1   4 p. (B)   TC 34 / SC 3*
Apple juice, apple juice concentrates and drinks containing apple
juice — Determination of patulin content —
Part 1: Method using high-performance liquid chromatography

ISO 8128-2:1993   *Ed. 1   5 p. (C)   TC 34 / SC 3*
Apple juice, apple juice concentrates and drinks containing apple
juice — Determination of patulin content —
Part 2: Method using thin-layer chromatography

67.180   Sugar. Sugar products. Starch

67.180.20 Starch and derived products

ISO 1227:1979   *Ed. 2   23 p. (L)   TC 93*
Starch, including derivatives and by-products — Vocabulary
Bilingual edition

ISO 1666:1973   *Ed. 1   6 p. (C)   TC 93*
Starch — Determination of moisture content — Oven-drying
methods

ISO/DIS 1666   *Ed. 1   5 p.   TC 93*
Starch — Determination of moisture content — Oven-drying
method (Revision of ISO 1666:1973)

ISO 1741:1980   *Ed. 1   2 p. (A)   TC 93*
Dextrose — Determination of loss in mass on drying — Vacuum
oven method

ISO 1742:1980   *Ed. 1   2 p. (A)   TC 93*
Glucose syrups — Determination of dry matter — Vacuum oven
method

ISO 1743:1982   *Ed. 2   11 p. (F)   TC 93*
Glucose syrup — Determination of dry matter
content — Refractive index method

ISO 3188:1978   *Ed. 1   3 p (B)   TC 93*
Starches and derived products — Determination of nitrogen
content by the Kjeldahl method — Titrimetric method

ISO 3593:1981   *Ed. 2   2 p. (A)   TC 93*
Starch — Determination of ash

ISO 3946:1982   *Ed. 1   3 p. (B)   TC 93*
Starches and derived products — Determination of total
phosphorus content — Spectrophotometric method

ISO 3947:1977   *Ed. 1   2 p. (A)   TC 93*
Starches, native or modified — Determination of total fat content

ISO 5377:1981   *Ed. 1   6 p. (C)   TC 93*
Starch hydrolysis products — Determination of reducing power
and dextrose equivalent — Lane and Eynon constant titre
method

ISO 5378:1978   *Ed. 1   3 p. (B)   TC 93*
Starches and derived products — Determination of nitrogen
content by the Kjeldahl method — Spectrophotometric method

ISO 5379:1983   *Ed. 1   6 p. (C)   TC 93*
Starches and derived products — Determination of sulfur dioxide
content — Acidimetric method and nephelometric method

ISO 5381:1983   *Ed. 1   6 p. (C)   TC 93 / SC 3*
Starch hydrolysis products — Determination of water
content — Modified Karl Fischer method

ISO 5809:1982   *Ed. 1   3 p. (B)   TC 93*
Starches and derived products — Determination of sulphated ash

ISO 5810:1982   *Ed. 1   2 p. (A)   TC 93*
Starches and derived products — Determination of chloride
content — Potentiometric method

ISO/DIS 10520   *Ed. 1   8 p.   TC 93*
Native starches — Determination of starch content — Ewers
polarimetric method

ISO/DIS 11212-1   *Ed. 1   7 p.   TC 93*
Starches and derived products — Determination of heavy metals
content —
Part 1: Determination of arsenic content by atomic absorption
spectrometry

ISO/DIS 11212-2   *Ed. 1   5 p.   TC 93*
Starches and derived products — Determination of heavy metals
content —
Part 2: Determination of mercury content by atomic absorption
spectrometry

ISO/DIS 11212-3   *Ed. 1   7 p.   TC 93*
Starches and derived products — Determination of heavy metals
content —
Part 3: Determination of lead content by atomic absorption
spectrometry with electrothermal atomization

ISO/DIS 11212-4   *Ed. 1   7 p.   TC 93*
Starches and derived products — Determination of heavy metals
content —
Part 4: Determination of cadmium content by atomic absorption
spectrometry with electrothermal atomization

ISO 11213:1995   *Ed. 1   8 p. (D)   TC 93*
Modified starch — Determination of acetyl content — Enzymatic
method

ISO/DIS 11214   *Ed. 1   5 p.   TC 93*
Modified starch — Determination of carboxyl group content of
oxidized starch

67.200 Edible oils and fats. Oilseeds

67.200.10 Animal and vegetable fats and oils

ISO 660:1983   *Ed. 1*   *4 p. (B)*   *TC 34 / SC 11*
Animal and vegetable fats and oils — Determination of acid value
and of acidity

ISO/DIS 660   *Ed. 2*   *TC 34 / SC 11*
Animal and vegetable fats and oils — Determination of acid value
and acidity (Revision of ISO 660:1983)

ISO 661:1989   *Ed. 2*   *2 p. (A)*   *TC 34 / SC 11*
Animal and vegetable fats and oils — Preparation of test sample

ISO 662:1980   *Ed. 1*   *2 p. (A)*   *TC 34 / SC 11*
Animal and vegetable fats and oils — Determination of moisture
and volatile matter content

ISO 663:1992   *Ed. 2*   *3 p. (B)*   *TC 34 / SC 11*
Animal and vegetable fats and oils — Determination of insoluble
impurities content

ISO 934:1980   *Ed. 1*   *2 p. (A)*   *TC 34 / SC 11*
Animal and vegetable fats and oils — Determination of water
content — Entrainment method

ISO 935:1988   *Ed. 1*   *3 p. (B)*   *TC 34 / SC 11*
Animal and vegetable fats and oils — Determination of titre

ISO 3596-1:1988   *Ed. 1*   *5 p. (C)*   *TC 34 / SC 11*
Animal and vegetable fats and oils — Determination of
unsaponifiable matter —
Part 1: Method using diethyl ether extraction (Reference method)

ISO 3596-2:1988   *Ed. 1*   *3 p. (B)*   *TC 34 / SC 11*
Animal and vegetable fats and oils — Determination of
unsaponifiable matter —
Part 2: Rapid method using hexane extraction

ISO 3656:1989   *Ed. 2*   *2 p. (A)*   *TC 34 / SC 11*
Animal and vegetable fats and oils — Determination of ultraviolet
absorbance

ISO 3657:1988   *Ed. 2*   *2 p. (A)*   *TC 34 / SC 11*
Animal and vegetable fats and oils — Determination of
saponification value

ISO 3960:1977   *Ed. 3*   *3 p. (B)*   *TC 34 / SC 11*
Animal and vegetable oils and fats — Determination of peroxide
value

ISO 3961:1989   *Ed. 2*   *2 p. (A)*   *TC 34 / SC 11*
Animal and vegetable fats and oils — Determination of iodine
value

ISO/DIS 3961   *Ed. 3*   *TC 34 / SC 11*
Animal and vegetable fats and oils — Determination of iodine
value (Revision of ISO 3961:1989)

ISO 5508:1990   *Ed. 2*   *7 p. (D)*   *TC 34 / SC 11*
Animal and vegetable fats and oils — Analysis by gas
chromatography of methyl esters of fatty acids

ISO 5509:1978   *Ed. 1*   *6 p. (C)*   *TC 34 / SC 11*
Animal and vegetable fats and oils — Preparation of methyl
esters of fatty acids

ISO 5555:1991   *Ed. 2*   *22 p. (L)*   *TC 34 / SC 11*
Animal and vegetable fats and oils — Sampling

ISO 5558:1982   *Ed. 1*   *3 p. (B)*   *TC 34 / SC 11*
Animal and vegetable fats and oils — Detection and identification
of antioxidants — Thin-layer chromatographic method

ISO 6320:1995   *Ed. 3*   *3 p. (B)*   *TC 34 / SC 11*
Animal and vegetable fats and oils — Determination of refractive
index

ISO 6321:1991   *Ed. 1*   *5 p. (C)*   *TC 34 / SC 11*
Animal and vegetable fats and oils — Determination of melting
point in open capillary tubes (slip point)

ISO 6463:1982   *Ed. 1*   *3 p. (B)*   *TC 34 / SC 11*
Animal and vegetable fats and oils — Determination of
butylhydroxyanisole (BHA) and butylhydroxytoluene
(BHT) — Gas-liquid chromatographic method

ISO 6464:1983   *Ed. 1*   *3 p. (B)*   *TC 34 / SC 11*
Animal and vegetable fats and oils — Determination of gallates
content — Molecular absorption spectrometric method

ISO 6656:1984   *Ed. 1*   *3 p. (B)*   *TC 34 / SC 11*
Animal and vegetable fats and oils — Determination of
polyethylene-type polymers

ISO 6799:1991   *Ed. 2*   *7 p. (D)*   *TC 34 / SC 11*
Animal and vegetable fats and oils — Determination of
composition of the sterol fraction — Method using gas
chromatography

ISO 6800:1985   *Ed. 1*   *5 p. (C)*   *TC 34 / SC 11*
Animal and vegetable fats and oils — Determination of the
composition of fatty acids in the 2-position

ISO 6883:1995   *Ed. 2*   *6 p. (C)*   *TC 34 / SC 11*
Animal and vegetable fats and oils — Determination of
conventional mass per volume ('litre weight in air')

ISO 6884:1985   *Ed. 1*   *2 p. (A)*   *TC 34 / SC 11*
Animal and vegetable fats and oils — Determination of ash

ISO 6885:1988   *Ed. 1*   *3 p. (B)*   *TC 34 / SC 11*
Animal and vegetable fats and oils — Determination of anisidine
value

ISO/DIS 6886   *Ed. 1*   *TC 34 / SC 11*
Animal and vegetable fats and oils — Determination of oxidation
stability (Accelerated oxidation test)

ISO 7366:1987   *Ed. 1*   *3 p. (B)*   *TC 34 / SC 11*
Animal and vegetable fats and oils — Determination of
1-monoglycerides and free glycerol contents

ISO 7847:1987   *Ed. 1*   *5 p. (C)*   *TC 34 / SC 11*
Animal and vegetable fats and oils — Determination of
polyunsaturated fatty acids with a cis,cis 1,4-diene structure

ISO 8292:1991   *Ed. 1*   *4 p. (C)*   *TC 34 / SC 11*
Animal and vegetable fats and oils — Determination of solid fat
content — Pulsed nuclear magnetic resonance method

ISO 8293:1990   *Ed. 1*   *6 p. (C)*   *TC 34 / SC 11*
Animal and vegetable fats and oils — Determination of dilatation

ISO 8294:1994   *Ed. 1*   *6 p. (C)*   *TC 34 / SC 11*
Animal and vegetable fats and oils — Determination of copper,
iron and nickel contents — Graphite furnace atomic absorption
method

ISO 8420:1990   *Ed. 1*   *5 p. (C)*   *TC 34 / SC 11*
Animal and vegetable fats and oils — Determination of polar
compounds content

ISO/DIS 8534   *Ed. 1*   *TC 34 / SC 11*
Animal and vegetable fats and oils — Determination of water
content — Karl Fischer method

ISO 9832:1992   *Ed. 1*   *5 p. (C)*   *TC 34 / SC 11*
Animal and vegetable fats and oils — Determination of residual
technical hexane content

ISO/DIS 9936   *Ed. 1*   *6 p.*   *TC 34 / SC 11*
Animal and vegetable fats and oils — Determination of
tocopherols and tocotrienols contents — Method by
high-performance liquid chromatography

ISO/DIS 10541   *Ed. 1*   *6 p.*   *TC 34 / SC 11*
Animal and vegetable fats and oils — Determination of the
absolute content of sterols — Enzymatic method

67.200.10

ISO 12193:1994   *Ed. 1   5 p. (C)*   *TC 34 / SC 11*
Animal and vegetable fats and oils — Determination of lead
content — Graphite furnace atomic absorption method

67.200.20 Oilseeds

ISO 542:1990   *Ed. 2   8 p. (D)*   *TC 34 / SC 2*
Oilseeds — Sampling

ISO 658:1988   *Ed. 2   4 p. (B)*   *TC 34 / SC 2*
Oilseeds — Determination of impurities content

ISO 659:1988   *Ed. 1   5 p. (C)*   *TC 34 / SC 2*
Oilseeds — Determination of hexane extract (or light petroleum
extract), called 'oil content'

ISO 664:1990   *Ed. 2   2 p. (B)*   *TC 34 / SC 2*
Oilseeds — Reduction of laboratory sample to test sample

ISO 665:1977   *Ed. 1   2 p. (A)*   *TC 34 / SC 2*
Oilseeds — Determination of moisture and volatile matter content

ISO 729:1988   *Ed. 2   3 p. (B)*   *TC 34 / SC 2*
Oilseeds — Determination of acidity of oils

ISO 734:1979   *Ed. 1   3 p. (B)*   *TC 34 / SC 2*
Oilseeds residues — Determination of hexane extract (or
light-petroleum extract), called 'oil content'

ISO 735:1977   *Ed. 1   2 p. (A)*   *TC 34 / SC 2*
Oilseed residues — Determination of ash insoluble in hydrochloric
acid

ISO 736:1977   *Ed. 1   3 p. (B)*   *TC 34 / SC 2*
Oilseed residues — Determination of diethyl ether extract

ISO 749:1977   *Ed. 1   2 p. (A)*   *TC 34 / SC 2*
Oilseed residues — Determination of total ash

ISO 771:1977   *Ed. 1   2 p. (A)*   *TC 34 / SC 2*
Oilseed residues — Determination of moisture and volatile matter
content

ISO 5500:1986   *Ed. 2   8 p. (D)*   *TC 34 / SC 2*
Oilseed residues — Sampling

ISO 5502:1992   *Ed. 2   4 p. (B)*   *TC 34 / SC 2*
Oilseed residues — Preparation of test samples

ISO 5504:1992   *Ed. 2   10 p. (E)*   *TC 34 / SC 2*
Oilseed residues — Determination of total isothiocyanate content
and vinylthiooxazolidone content

ISO 5506:1988   *Ed. 2   2 p. (A)*   *TC 34 / SC 2*
Soya bean products — Determination of urease activity

ISO 5507:1992   *Ed. 2   13 p. (G)*   *TC 34 / SC 2*
Oilseeds — Nomenclature
Trilingual edition

ISO 5511:1992   *Ed. 2   5 p. (C)*   *TC 34 / SC 2*
Oilseeds — Determination of oil content — Method using
continuous-wave low-resolution nuclear magnetic resonance
spectrometry (Rapid method)

ISO 5512:1982   *Ed. 1   1 p. (A)*   *TC 34 / SC 2*
Sunflower seed for the manufacture of oil — Specification

ISO 5514:1979   *Ed. 1   2 p. (A)*   *TC 34 / SC 2*
Soya bean products — Determination of cresol red index

ISO 8892:1987   *Ed. 1   3 p. (B)*   *TC 34 / SC 2*
Oilseed residues — Determination of total residual hexane

ISO 9167-1:1992   *Ed. 1   9 p. (E)*   *TC 34 / SC 2*
Rapeseed — Determination of glucosinolates content —
Part 1: Method using high-performance liquid chromatography

ISO 9167-2:1994   *Ed. 1   6 p. (C)*   *TC 34 / SC 2*
Rapeseed — Determination of glucosinolates content —
Part 2: Method using X-ray fluorescence spectrometry

ISO 9289:1991   *Ed. 1   4 p. (B)*   *TC 34 / SC 2*
Oilseed residues — Determination of free residual hexane

ISO 10519:1992   *Ed. 1   8 p. (D)*   *TC 34 / SC 2*
Rapeseed — Determination of chlorophyll
content — Spectrometric method

ISO/DIS 10519   *Ed. 1   11 p.*   *TC 34 / SC 2*
Rapeseed — Determination of chlorophyll
content — Spectrometric method (Revision of ISO 10519:1992)

ISO 10565:1993   *Ed. 1   7 p. (D)*   *TC 34 / SC 2*
Oilseeds — Simultaneous determination of oil and moisture
contents — Method using pulsed nuclear magnetic resonance
spectrometry

ISO 10633-1:1995   *Ed. 1   9 p. (E)*   *TC 34 / SC 2*
Oilseed residues — Determination of glucosinolates content —
Part 1: Method using high-performance liquid chromatography

67.220   Spices and condiments. Food additives

67.220.10 Spices and condiments

ISO 676:1995   *Ed. 2   25 p. (L)*   *TC 34 / SC 7*
Spices and condiments — Botanical nomenclature
Bilingual edition

ISO 882-1:1993   *Ed. 1   9 p. (E)*   *TC 34 / SC 7*
Cardamom (Elettaria cardamomum (Linnaeus) Maton var.
minuscula Burkill) — Specification —
Part 1: Whole capsules

     Draft Technical Corrigendum 1 to ISO 882-1:1993
     Technical Corrigendum 1
                              Ed. 1   *TC 34 / SC 7*

ISO 882-2:1993   *Ed. 1   4 p. (B)*   *TC 34 / SC 7*
Cardamom (Elettaria cardamomum (Linnaeus) Maton var.
minuscula Burkill) — Specification —
Part 2: Seeds

     Draft Technical Corrigendum 1 to ISO 882-2:1993
     Technical Corrigendum 1
                              Ed. 1   *TC 34 / SC 7*

ISO 927:1982   *Ed. 2   1 p. (A)*   *TC 34 / SC 7*
Spices and condiments — Determination of extraneous matter
content

ISO 928:1980   *Ed. 1   2 p. (A)*   *TC 34 / SC 7*
Spices and condiments — Determination of total ash

ISO/DIS 928   *Ed. 1   5 p.*   *TC 34 / SC 7*
Spices and condiments — Determination of total ash (Revision of
ISO 928:1980)

ISO 930:1980   *Ed. 1   2 p. (A)*   *TC 34 / SC 7*
Spices and condiments — Determination of acid-insoluble ash

ISO/DIS 930   *Ed. 1   5 p.*   *TC 34 / SC 7*
Spices and condiments — Determination of acid-insoluble ash
(Revision of ISO 930:1980)

ISO 939:1980   *Ed. 1   4 p. (B)*   *TC 34 / SC 7*
Spices and condiments — Determination of moisture
content — Entrainment method

ISO 941:1980   *Ed. 1   2 p. (A)*   *TC 34 / SC 7*
Spices and condiments — Determination of cold water-soluble
extract

ISO 948:1980 *Ed. 1* *2 p. (A)* *TC 34 / SC 7*
Spices and condiments — Sampling

ISO 959-1:1989 *Ed. 1* *11 p. (F)* *TC 34 / SC 7*
Pepper (Piper nigrum Linnaeus), whole or
ground — Specification —
Part 1: Black pepper

ISO 959-2:1989 *Ed. 1* *8 p. (D)* *TC 34 / SC 7*
Pepper (Piper nigrum Linnaeus), whole or
ground — Specification —
Part 2: White pepper

ISO 972:1985 *Ed. 1* *5 p. (C)* *TC 34 / SC 7*
Chillies and capsicums, whole or ground
(powdered) — Specification

ISO/DIS 972 *Ed. 2* *6 p.* *TC 34 / SC 7*
Chillies and capsicums, whole or ground
(powdered) — Specification (Revision of ISO 972:1985)

ISO 973:1980 *Ed. 1* *4 p. (B)* *TC 34 / SC 7*
Spices and condiments — Pimento (allspice), whole or
ground — Specification

ISO 1003:1980 *Ed. 1* *6 p. (C)* *TC 34 / SC 7*
Spices and condiments — Ginger, whole, in pieces, or
ground — Specification

ISO 1108:1992 *Ed. 2* *2 p. (A)* *TC 34 / SC 7*
Spices and condiments — Determination of non-volatile ether
extract

ISO 1208:1982 *Ed. 1* *5 p. (C)* *TC 34 / SC 7*
Spices and condiments — Determination of filth

ISO 1237:1981 *Ed. 2* *11 p. (F)* *TC 34 / SC 7*
Mustard seed — Specification

ISO 2253:1986 *Ed. 2* *3 p. (B)* *TC 34 / SC 7*
Curry powder — Specification

ISO/DIS 2253 *Ed. 1* *13 p.* *TC 34 / SC 7*
Curry powder — Specification (Revision of ISO 2253:1986)

ISO 2254:1980 *Ed. 1* *4 p. (B)* *TC 34 / SC 7*
Cloves, whole and ground (powdered) — Specification

ISO 2255:1980 *Ed. 1* *4 p. (B)* *TC 34 / SC 7*
Coriander, whole or ground (Powdered) — Specification

ISO/DIS 2255 *Ed. 2* *4 p.* *TC 34 / SC 7*
Coriander (Coriandrum sativum L.), whole or ground
(powdered) — Specification (Revision of ISO 2255:1980)

ISO 2256:1984 *Ed. 1* *3 p. (B)* *TC 34 / SC 7*
Dried mint (spearmint) (Mentha spicata Linnaeus syn. Mentha
viridis Linnaeus) — Specification

ISO 2825:1981 *Ed. 2* *1 p. (A)* *TC 34 / SC 7*
Spices and condiments — Preparation of a ground sample for
analysis

ISO 3493:1976 *Ed. 1* *2 p. (A)* *TC 34 / SC 7*
Vanilla — Vocabulary
Bilingual edition

ISO 3513:1995 *Ed. 2* *7 p. (D)* *TC 34 / SC 7*
Chillies — Determination of Scoville index

ISO 3588:1977 *Ed. 2* *2 p. (A)* *TC 34 / SC 7*
Spices and condiments — Determination of degree of fineness of
grinding — Hand sieving method (Reference method)

ISO 3632-1:1993 *Ed. 1* *6 p. (C)* *TC 34 / SC 7*
Saffron (Crocus sativus Linnaeus) —
Part 1: Specification

ISO 3632-2:1993 *Ed. 1* *12 p. (F)* *TC 34 / SC 7*
Saffron (Crocus sativus Linnaeus) —
Part 2: Test methods

ISO 5559:1995 *Ed. 3* *9 p. (E)* *TC 34 / SC 7*
Dehydrated onion (Allium cepa Linnaeus) — Specification

ISO 5560:1983 *Ed. 2* *6 p. (C)* *TC 34 / SC 7*
Dehydrated garlic — Specification

ISO/DIS 5560 *Ed. 3* *10 p.* *TC 34 / SC 7*
Dehydrated garlic (Allium sativum L.) — Specification (Revision of
ISO 5560:1983)

ISO 5561:1990 *Ed. 2* *4 p. (B)* *TC 34 / SC 7*
Black caraway and blond caraway (Carum carvi Linnaeus),
whole — Specification

ISO 5562:1983 *Ed. 1* *3 p. (B)* *TC 34 / SC 7*
Turmeric, whole or ground (powdered) — Specification

ISO 5563:1984 *Ed. 1* *5 p. (C)* *TC 34 / SC 7*
Dried peppermint (Mentha piperita Linnaeus) — Specification

ISO 5564:1982 *Ed. 1* *2 p. (A)* *TC 34 / SC 7*
Black pepper and white pepper, whole or
ground — Determination of piperine content —
Spectrophotometric method

ISO 5565:1982 *Ed. 1* *8 p. (D)* *TC 34 / SC 7*
Vanilla (Vanilla fragrans (Salisbury) Ames) — Specification

ISO 5566:1982 *Ed. 1* *2 p. (A)* *TC 34 / SC 7*
Turmeric — Determination of colouring
power — Spectrophotometric method

ISO 5567:1982 *Ed. 1* *3 p. (B)* *TC 34 / SC 7*
Dehydrated garlic — Determination of volatile organic sulphur
compounds

ISO 6465:1984 *Ed. 1* *3 p. (J)* *TC 34 / SC 7*
Whole cumin (Cuminum cyminum Linnaeus) — Specification

ISO 6538:1982 *Ed. 1* *4 p. (B)* *TC 34 / SC 7*
Cassia (type China, type Indonesia and type Viet Nam), whole and
ground (powdered) — Specification

ISO/DIS 6538 *Ed. 1* *7 p.* *TC 34 / SC 7*
Cassia, Chinese type, Indonesian type and Vietnamese type
Cinnamomum aromaticum (Nees) syn. Cinnamomum cassia
(Nees) ex. Blume, Cinnamomum burmanii (C.G. Nees) Blume and
Cinnamomum loureirii Nees" — Specification (Revision of ISO
6538:1982)

ISO 6539:1983 *Ed. 1* *6 p. (C)* *TC 34 / SC 7*
Cinnamon (type Sri Lanka (Ceylon), type Seychelles and type
Madagascar), whole or ground (powdered) — Specification

ISO/DIS 6539 *Ed. 1* *8 p.* *TC 34 / SC 7*
Cinnamon, Sri Lankan type, Seychelles type and Madagascan type
(Cinnamomum zeylanicum Blume) — Specification (Revision of
ISO 6539:1983)

ISO 6571:1984 *Ed. 1* *4 p. (B)* *TC 34 / SC 7*
Spices, condiments and herbs — Determination of volatile oil
content

ISO 6574:1986 *Ed. 1* *3 p. (B)* *TC 34 / SC 7*
Celery seed (Apium graveolens Linnaeus) — Specification

ISO 6575:1982 *Ed. 1* *3 p. (B)* *TC 34 / SC 7*
Fenugreek, whole or ground (powdered) — Specification

ISO 6576:1984 *Ed. 1* *4 p. (B)* *TC 34 / SC 7*
Laurel (Laurus nobilis Linnaeus) — Whole and pounded
leaves — Specification

ISO 6577:1990 *Ed. 1* *7 p. (D)* *TC 34 / SC 7*
Nutmeg, whole or broken, and mace, whole or in pieces (Myristica
fragrans Houttuyn) — Specification

ISO 6754:1985 *Ed. 1* *4 p. (B)* *TC 34 / SC 7*
Whole thyme (Thymus vulgaris Linnaeus) — Specification

ISO/DIS 6754 *Ed. 2* *4 p.* *TC 34 / SC 7*
Dried thyme (Thymus vulgaris L.) — Specification (Revision of
ISO 6754:1985)

67.220.10

ISO 7377:1984　　　　*Ed. 1　　3 p. (B)*　　　*TC 34 / SC 7*
Juniper berries (Juniperus communis Linnaeus) — Specification

ISO 7386:1984　　　　*Ed. 1　　3 p. (B)*　　　*TC 34 / SC 7*
Aniseed (Pimpinella anisum Linnaeus) — Specification

ISO 7540:1984　　　　*Ed. 1　　4 p. (B)*　　　*TC 34 / SC 7*
Ground (powdered) paprika (Capsicum annuum
Linnaeus) — Specification

ISO 7541:1989　　　　*Ed. 1　　3 p. (B)*　　　*TC 34 / SC 7*
Ground (powdered) paprika — Determination of total natural
colouring matter content

ISO 7542:1984　　　　*Ed. 1　　6 p. (C)*　　　*TC 34 / SC 7*
Ground (powdered) paprika (Capsicum annuum
Linnaeus) — Microscopical examination

ISO 7543-1:1994　　　　*Ed. 1　　4 p. (B)*　　　*TC 34 / SC 7*
Chillies and chilli oleoresins — Determination of total capsaicinoid
content —
Part 1: Spectrometric method

ISO 7543-2:1993　　　　*Ed. 1　　5 p. (C)*　　　*TC 34 / SC 7*
Chillies and chilli oleoresins — Determination of total capsaicinoid
content —
Part 2: Method using high-performance liquid chromatography

ISO 7925:1985　　　　*Ed. 1　　3 p. (B)*　　　*TC 34 / SC 7*
Dried oregano (Origanum vulgare Linnaeus) — Whole or ground
leaves — Specification

ISO 7926:1991　　　　*Ed. 1　　7 p. (D)*　　　*TC 34 / SC 7*
Dehydrated tarragon (Artemisia dracunculus
Linnaeus) — Specification

ISO 7927-1:1987　　　　*Ed. 1　　4 p. (B)*　　　*TC 34 / SC 7*
Fennel seed, whole or ground (powdered) —
Part 1: Bitter fennel seed (Foeniculum vulgare P. Miller var.
vulgare) — Specification

ISO 7928-1:1991　　　　*Ed. 1　　7 p. (D)*　　　*TC 34 / SC 7*
Savory — Specification —
Part 1: Winter savory (Satureja montana Linnaeus)

ISO 7928-2:1991　　　　*Ed. 1　　7 p. (D)*　　　*TC 34 / SC 7*
Savory — Specification —
Part 2: Summer savory (Satureja hortensis Linnaeus)

ISO 10620:1995　　　　*Ed. 1　　4 p. (B)*　　　*TC 34 / SC 7*
Dried sweet marjoram (Origanum majorana L.) — Specification

ISO/DIS 10621　　　　*Ed. 1　　5 p.*　　　*TC 34 / SC 7*
Dehydrated green pepper (Piper nigrum L.) — Specification

ISO/DIS 10622　　　　*Ed. 1　　6 p.*　　　*TC 34 / SC 7*
Large cardamom (Amomum subulatum Roxburgh), as capsules
and seeds — Specification

ISO 11027:1993　　　　*Ed. 1　　5 p. (C)*　　　*TC 34 / SC 7*
Pepper and pepper oleoresins — Determination of piperine
content — Method using high-performance liquid chromatography

ISO 11163:1995　　　　*Ed. 1　　3 p. (B)*　　　*TC 34 / SC 7*
Dried sweet basil (Ocimum basilicum L.) — Specification

ISO 11164:1995　　　　*Ed. 1　　3 p. (B)*　　　*TC 34 / SC 7*
Dried rosemary (Rosmarinus officinalis L.) — Specification

ISO 11165:1995　　　　*Ed. 1　　3 p. (B)*　　　*TC 34 / SC 7*
Dried sage (Salvia officinalis L.) — Specification

ISO 11178:1995　　　　*Ed. 1　　7 p. (D)*　　　*TC 34 / SC 7*
Star anise (Illicium verum Hook. f.) — Specification

ISO/DIS 13685　　　　*Ed. 1　　13 p.*　　　*TC 34 / SC 7*
Ginger and its oleoresins — Determination of the main pungent
components (gingerols and shogaols) — Method using
high-performance liquid chromatography

67.220.20　Food additives

ISO 2515:1973　　　　*Ed. 1　　3 p. (B)*　　　*TC 47*
Ammonium hydrogen carbonate for industrial use (including
foodstuffs) — Determination of ammoniacal nitrogen
content — Volumetric method after distillation

ISO 2516:1973　　　　*Ed. 1　　1 p. (A)*　　　*TC 47*
Ammonium hydrogen carbonate for industrial use (including
foodstuffs) — Determination of total alkalinity — Volumetric
method

ISO 3360:1976　　　　*Ed. 1　　5 p. (C)*　　　*TC 47*
Phosphoric acid and sodium phosphates for industrial use
(including foodstuffs) — Determination of fluorine
content — Alizarin complexone and lanthanum nitrate
photometric method

ISO 3420:1975　　　　*Ed. 1　　1 p. (A)*　　　*TC 47*
Ammonium hydrogen carbonate for industrial use (including
foodstuffs) — Determination of ash — Gravimetric method

ISO 3422:1975　　　　*Ed. 1　　3 p. (B)*　　　*TC 47*
Ammonium hydrogen carbonate for industrial use (including
foodstuffs) — Determination of total carbon dioxide
content — Titrimetric method

ISO 3706:1976　　　　*Ed. 1　　3 p. (B)*　　　*TC 47*
Phosphoric acid for industrial use (including
foodstuffs) — Determination of total phosphorus(V) oxide
content — Quinoline phosphomolybdate gravimetric method

ISO 3707:1976　　　　*Ed. 1　　4 p. (B)*　　　*TC 47*
Phosphoric acid for industrial use (including
foodstuffs) — Determination of calcium content — Flame atomic
absorption method

ISO 3708:1976　　　　*Ed. 1　　4 p. (B)*　　　*TC 47*
Phosphoric acid for industrial use (including
foodstuffs) — Determination of chloride
content — Potentiometric method

ISO 3709:1976　　　　*Ed. 1　　4 p. (B)*　　　*TC 47*
Phosphoric acid for industrial use (including
foodstuffs) — Determination of oxides of nitrogen
content — 3,4- Xylenol spectrophotometric method

ISO 4275:1977　　　　*Ed. 1　　2 p. (A)*　　　*TC 47*
Ammonium hydrogen carbonate for industrial use (including
foodstuffs) — Determination of arsenic content — Silver
diethyldithiocarbamate photometric method

ISO 5372:1978　　　　*Ed. 1　　2 p. (A)*　　　*TC 47*
Condensed phosphates for industrial use (including
foodstuffs) — Determination of arsenic content — Silver
diethyldithiocarbamate photometric method

ISO 5373:1981　　　　*Ed. 1　　4 p. (B)*　　　*TC 47*
Condensed phosphates for industrial use (including
foodstuffs) — Determination of calcium content — Flame atomic
absorption spectrometric method

ISO 5374:1978　　　　*Ed. 1　　4 p. (B)*　　　*TC 47*
Condensed phosphates for industrial use (including
foodstuffs) — Determination of chloride
content — Potentiometric method

ISO 5375:1979　　　　*Ed. 1　　4 p. (B)*　　　*TC 47*
Condensed phosphates for industrial use (including
foodstuffs) — Determination of oxides of nitrogen
content — 3,4- Xylenol spectrophotometric method

ISO 7099:1983　　　　*Ed. 1　　4 p. (B)*　　　*TC 47*
Phosphoric acid for industrial use (including
foodstuffs) — Determination of hydrogen sulfide
content — Titrimetric method

ISO 7100:1983　　　　*Ed. 1　　2 p. (A)*　　　*TC 47*
Phosphoric acid for industrial use (including
foodstuffs) — Determination of vanadium
content — Phosphotungsten vanadate spectrometric method

ISO 7110:1985   *Ed. 1   3 p. (B)   TC 47*
Ammonium bicarbonate (Ammonium hydrogen- carbonate) for industrial use (including foodstuffs) — Determination of lead content — Flame atomic absorption method

67.240   Sensory analysis

ISO 3591:1977   *Ed. 1   3 p. (B)   TC 34 / SC 12*
Sensory analysis — Apparatus — Wine-tasting glass

ISO 3972:1991   *Ed. 2   7 p. (D)   TC 34 / SC 12*
Sensory analysis — Methodology — Method of investigating sensitivity of taste

ISO 4120:1983   *Ed. 1   8 p. (D)   TC 34 / SC 12*
Sensory analysis — Methodology — Triangular test

ISO 4121:1987   *Ed. 1   7 p. (D)   TC 34 / SC 12*
Sensory analysis — Methodology — Evaluation of food products by methods using scales

ISO 5492:1992   *Ed. 1   22 p. (L)   TC 34 / SC 12*
Sensory analysis — Vocabulary
Bilingual edition

ISO 5494:1978   *Ed. 1   3 p. (B)   TC 34 / SC 12*
Sensory analysis — Apparatus — Tasting glass for liquid products

ISO 5495:1983   *Ed. 2   6 p. (C)   TC 34 / SC 12*
Sensory analysis — Methodology — Paired comparison test

ISO 5496:1992   *Ed. 1   14 p. (G)   TC 34 / SC 12*
Sensory analysis — Methodology — Initiation and training of assessors in the detection and recognition of odours

ISO 5497:1982   *Ed. 1   2 p. (A)   TC 34 / SC 12*
Sensory analysis — Methodology — Guidelines for the preparation of samples for which direct sensory analysis is not feasible

ISO 6564:1985   *Ed. 1   6 p. (C)   TC 34 / SC 12*
Sensory analysis — Methodology — Flavour profile methods

ISO 6658:1985   *Ed. 1   14 p. (G)   TC 34 / SC 12*
Sensory analysis — Methodology — General guidance

ISO 6668:1991   *Ed. 1   3 p. (B)   TC 34 / SC 15*
Green coffee — Preparation of samples for use in sensory analysis

ISO 8586-1:1993   *Ed. 1   15 p. (H)   TC 34 / SC 12*
Sensory analysis — General guidance for the selection, training and monitoring of assessors —
Part 1: Selected assessors

ISO 8586-2:1994   *Ed. 1   10 p. (E)   TC 34 / SC 12*
Sensory analysis — General guidance for the selection, training and monitoring of assessors —
Part 2: Experts

ISO 8587:1988   *Ed. 1   9 p. (E)   TC 34 / SC 12*
Sensory analysis — Methodology — Ranking

ISO 8588:1987   *Ed. 1   6 p. (C)   TC 34 / SC 12*
Sensory analysis — Methodology — 'A' - 'not A' test

ISO 8589:1988   *Ed. 1   9 p. (E)   TC 34 / SC 12*
Sensory analysis — General guidance for the design of test rooms

ISO 10399:1991   *Ed. 1   6 p. (C)   TC 34 / SC 12*
Sensory analysis — Methodology — Duo-trio test

ISO 11035:1994   *Ed. 1   26 p. (M)   TC 34 / SC 12*
Sensory analysis — Identification and selection of descriptors for establishing a sensory profile by a multidimensional approach

ISO 11036:1994   *Ed. 1   14 p. (G)   TC 34 / SC 12*
Sensory analysis — Methodology — Texture profile

ISO/DIS 11037   *Ed. 1   8 p.   TC 34 / SC 12*
Sensory analysis — General guidance and test method for the assessment of the colour of foods

ISO/DIS 11056   *Ed. 1   25 p.   TC 34 / SC 12*
Sensory analysis — Methodology — Magnitude estimation

ISO/DIS 13299   *Ed. 1   38 p.   TC 34 / SC 12*
Sensory analysis — Methodology — General guidance for establishing a sensory profile

ISO/DIS 13301   *Ed. 1   19 p.   TC 34 / SC 12*
Sensory analysis — Methodology — General guidance for defining and calculating individual and group sensory thresholds from three alternative forced-choice data sets

67.260   Plants and equipment for the food industry

ISO 488:1983   *Ed. 1   10 p. (E)   TC 34 / SC 5*
Milk — Determination of fat content — Gerber butyrometers

ISO 2037:1992   *Ed. 1   3 p. (B)   TC 5 / SC 1*
Stainless steel tubes for the food industry

ISO 2449:1974   *Ed. 1   8 p. (D)   TC 34 / SC 5*
Milk and liquid milk products — Density hydrometers for use in products with a surface tension of approximately 45 mN/m

ISO 2851:1993   *Ed. 1   6 p. (C)   TC 5 / SC 1*
Stainless steel bends and tees for the food industry

ISO 2852:1993   *Ed. 2   12 p. (F)   TC 5 / SC 1*
Stainless steel clamp pipe couplings for the food industry

ISO 2853:1993   *Ed. 2   23 p. (L)   TC 5 / SC 1*
Stainless steel threaded couplings for the food industry

ISO 3432:1975   *Ed. 1   4 p. (B)   TC 34 / SC 5*
Cheese — Determination of fat content — Butyrometer for Van Gulik method

ISO 3591:1977   *Ed. 1   3 p. (B)   TC 34 / SC 12*
Sensory analysis — Apparatus — Wine-tasting glass

ISO 3889:1977   *Ed. 1   3 p. (B)   TC 34 / SC 5*
Milk and milk products — Determination of fat content — Mojonnier-type fat extraction flasks

ISO 5223:1995   *Ed. 3   4 p. (B)   TC 34 / SC 4*
Test sieves for cereals

ISO 5494:1978   *Ed. 1   3 p. (B)   TC 34 / SC 12*
Sensory analysis — Apparatus — Tasting glass for liquid products

ISO 6666:1983   *Ed. 1   2 p. (A)   TC 34 / SC 15*
Coffee triers

ISO 7700-1:1984   *Ed. 1   6 p. (C)   TC 34 / SC 4*
Check of the calibration of moisture meters —
Part 1: Moisture meters for cereals

ISO 7700-2:1987   *Ed. 1   6 p. (C)   TC 34 / SC 2*
Check of the calibration of moisture meters —
Part 2: Moisture meters for oilseeds

81   GLASS AND CERAMICS INDUSTRIES

81.060   Ceramics

81.060.20 Ceramic products

ISO/TTA 1:1994   *Ed. 1   62 p. (R)   VAMAS*
Advanced technical ceramics — Unified classification system

ISO 6486-1:1981   *Ed. 1   4 p. (B)   TC 166*
Ceramic ware in contact with food — Release of lead and
cadmium —
Part 1: Method of test

ISO 6486-2:1981   *Ed. 1   2 p. (A)   TC 166*
Ceramic ware in contact with food — Release of lead and
cadmium —
Part 2: Permissible limits

ISO 7086-1:1982   *Ed. 1   5 p. (C)   TC 166 / SC 2*
Glassware and glass ceramic ware in contact with
food — Release of lead and cadmium —
Part 1: Method of test

ISO 7086-2:1982   *Ed. 1   1 p. (A)   TC 166 / SC 2*
Glassware and glass ceramic ware in contact with
food — Release of lead and cadmium —
Part 2: Permissible limits

ISO 8391-1:1986   *Ed. 1   4 p. (B)   TC 166 / SC 1*
Ceramic cookware in contact with food — Release of lead and
cadmium —
Part 1: Method of test

ISO 8391-2:1986   *Ed. 1   2 p. (A)   TC 166 / SC 1*
Ceramic cookware in contact with food — Release of lead and
cadmium —
Part 2: Permissible limits

ISO CENTRAL SECRETARIAT

1, rue de Varembé
Case postale 56
CH-1211 Genève 20
Switzerland
Telephone: + 41 22 749 01 11

Sales Department

Telefax: + 41 22 734 10 79
Email:
    Internet: sales@isocs.iso.ch
    X.400:   c=ch; a=400net; p=iso; o=isocs; s=sales

Addresses of Authors

Joe O.K. Boison
Health of Animals Laboratory
Agriculture & Agri-Food Canada
116 Veterinary Road
Saskatoon
Saskatchewan
CANADA S7N 2R3

A. R. Byrne
J. Stefan Institute
61111 Ljubljana
SLOVENIA

Stephen L R Ellison
Laboratory of the Government Chemist
Queens Road
Teddington
Middlesex
ENGLAND TW11 OLY

William Horwitz
U.S. Food and Drug Administration
HFS-500
200 C Street SW
Washington
D.C. 20204
USA

A. Jaksakul
Dpt of Environmental Quality Promotion
Environmental Research and Training
center, Technopolis
Klong Luang District
Pathumthani 12120
THAILAND

R. Boonyatumanond
Dpt of Environmental Quality Promotion
Environmental Research and Training
center, Technopolis
Klong Luang District
Pathumthani 12120
THAILAND

Robert W. Dabeka
Food Research Division
Health Protection Branch
Health Canada
Ottawa, Ontario K1A 0L2
CANADA

A. Fajgelj
International Atomic Energy Agency
Laboratories
A-2444 Seibersdorf
AUSTRIA

Milan Ihnat
Centre for Land and Biological Resources
Research, Research Branch
Agriculture and Agri-Food Canada
Ottawa
Ontario K1A 0C6
CANADA

J. F. Kay
Veterinary Medicines Directorate
Woodham Lane
New Haw,
Addlestone
Surrey
U.K. KT15 3NB

Del A. Koch
ABC Laboratories
7200 E. ABC Lane
Columbia
MO 65202
USA

James D. MacNeil
Health of Animals Laboratory
Agriculture & Agri-Food Canada
116 Veterinary Road
Saskatoon
Saskatchewan
CANADA S7N 2R3

Valerie K. Martz
Health of Animals Laboratory
Agriculture & Agri-Food Canada
116 Veterinary Road
Saskatoon
Saskatchewan
CANADA S7N 2R3

Patrick A. Noland
ABC Laboratories
7200 E. ABC Lane
Columbia
MO 65202
USA

P. Prinyatanakun
Dpt of Environmental Quality Promotion
Environmental Research and Training
center, Technopolis
Klong Luang District
Pathumthani 12120
THAILAND

Loren C. Schrier
ABC Laboratories
7200 E. ABC Lane
Columbia
MO 65202
USA

P. Prinyatanakun
Dpt of Environmental Quality Promotion
Environmental Research and Training
center, Technopolis
Klong Luang District
Pathumthani 12120
THAILAND

Alex Williams
19 Hamesmoor Way
Mytchett
Camberley
Surrey
ENGLAND GU16 6JG

Subject Index

A

ABC Laboratories 38
Acceptable Daily Intake (ADI)..43; 61
acceptable performance 30
accepted results 18
accumulated error..17; 18
accuracy..10; 20; 25; 27; 59
activated radioisotope..28
added analyte 8; 9; 10; 15; 18; 22
additional factor 3
additive factor 4
aldrin..48
amount of analyte..6; 35
analyst training..8
analyte 1
analyte concentration..14
analyte levels..12
analyte present 9
analyte recovery..5; 43; 44
analyte volatilization...7
analytical blank..24
analytical data 30
analytical system 8
approximations...32
arsenic..26
assay..41
associated uncertainties 16
associated uncertainty..36
asymmetric distribution 35
averages 2

B

baseline..14
baseline errors..15
baseline level 13; 18
baseline noise...20
baseline value...13
behavioral differences..7
best estimate 2; 35
best estimate concentrations 16
best estimate values...18
bias 1; 3; 30; 33; 36; 41
biases 19
bioaccumulation 48
biological matrices 42
biological systems 4
blank...1; 4; 14; 19; 20; 21; 32
blank control 14
blank determinations 14
blank effects...14
blank matrix...2
blank sample...46
blank value...14
blended recoveries..15

bracketing 6

C

cadmium...26
calculation...6
calculation of recovery...13
calibrants 9
calibration 6
calibration curve..6; 46
calibration curve fitting 6
calibration solutions 6; 9
calibration technique 6
carrier...25
ceramic ware..32
certificate 12; 13; 16
certificate of analysis...18; 19
certification of reference materials 28
certified limit..40
Certified Reference Material 12
certified value 11; 19; 33; 34
certifying body...12
chemical measurement 31
chemical yield..25
chemisorption...11
chromatographic separation 60
chromatographic software...20
coal fly ash...28
cobalt...26
coefficient of variation..28
collection techniques 7
composite stock solution..9
concentration values...12
concept of uncertainty..31
confidence interval 19
confidence intervals...30
consensus for harmonization 5
constant error 1; 3
constant recovery factor 3
contaminant..14
contamination..6; 7; 21; 25; 32
control...1; 4
control charts..46
control material 11; 18; 21
copper 26
corrected residue values...39
corrected result..1; 3
correction 21; 34; 35; 41
correction factor 1; 3; 4; 39
correction for recovery 40; 46
correction of analytical results 22
coverage factor...34
critical value...35
cross contamination 32
cut-off point...20

D

data handling..6
data interpretation 6

data presentation 6
data reduction algorithm 32
DDTs 48
dead time 25
dependent variables...17
detection limit 20; 48; 56
determination of recovery 9; 12
determination of uncertainty...30
diagnostic steps 21
dieldrin 48
diethylstilbesterol (DES)...45
direct calibrant...27
discrimination threshold 32
dissipation studies..40
drug residue...42

E

emission spectrometers...20
empirical method...4; 32; 35; 36
endogenous analyte...5; 15
endrin...48
enforcement limit 7
environmental conditions...32
environmental factors 6
environmental fate studies...40
environmental Reference Materials...11
environmental samples 27
error...18; 21
estimate of recovery..8
estimate of uncertainty...34; 35
estimated measurement uncertainty..36
estimated recovery...34; 36
estimating recovery...10; 22
estimation of recovery..5; 46
Eurachem 30
Eurachem guide...31
European Commission Decision 61
European Union regulations...61
evaluation of accuracy 21
evaluation of uncertainty...24
evaporation...42
expanded uncertainty...31; 34
experimental matrix 33
external standard...46; 60
extraction 3; 7; 32; 33; 44
extraction recovery 31
extraction time 54

F

Federal Insecticide, Fungicide, and Rodenticide Act (FIFRA). 38
final recovery factor...17; 18
fixed recovery factor 35
Food and Drug Administration (FDA)..38
food chain 48
food composition specifications 4
formulation analysis 40
fortified samples 38; 39; 46

G

gamma energy 26
gamma-rays...26
Ge detectors 26
genetic predisposition...6
government regulations...5
gross analyte value 14
groundwater studies 40

H

half-life..26
harmonization..28
HCHs..48
health hazards 48
homogeneity..17; 21

I

identifiable peaks..20
identification software...20
imperfect recovery...35; 36
imprecision 21
incomplete sample dissolution..11
incurred residue..4
independent method 19; 21
independent variables..17
indicator radionuclides IRN..26
individual recovery factor 8
in-house quality control 13
in-house quality control materials 11
injection error..49
inorganic elemental analytes...5
input values..3
insoluble siliceous residues 11
instrument calibration..35
instrument resolution...32
instrument response...33
instrumental neutron activation analysis INAA...24; 25
interactions...10; 21
intercomparison..28; 33
interference 6; 7; 10; 24; 32
interlaboratory comparisons..30
interlaboratory systematic errors 18
interlaboratory testing...44
intermediate parameters 31
internal standard..3; 4; 45; 48; 49; 51; 60
internal standardization 22
international trade 5
intrinsic errors...6
invalid baseline 21
invalid blank 20
iodine..26
irradiation standard 24; 27
ISO - International Organisation for Standardization 30
ISO Guide 31
isotope 25; 28
isotope labelling techniques 45
isotopic exchange 25

L

labelled molecule 60
laboratory bias... 5
laboratory imprecisions.. 19
laboratory performance 5; 43
laboratory sample 6
law of propagation of error 1; 3
lead 21; 26; 32
legal compliance.. 5
legal mandate 12
level of confidence 31; 34
limit of determination 7
limit of quantitation 7
limit of reporting 7
low-energy photon detectors.. 26

M

magnitude of errors 14
magnitude of the residue (MOR) 38
marginal recovery.. 4
marker residue 61
mass spectral assays.. 45
mass spectrometer... 45
mass spectrometry 60
material homogeneity 18
material inhomogeneity 19
matrix 4; 24; 28; 32; 49
matrix blank.. 14
matrix composition.. 9; 12
matrix effect.. 6; 31
Maximum Residue Limit (MRL).. 43; 61
mean recovery 34
measurand.. 1; 31; 32; 33; 35
measured recovery.. 8
measured value... 1
measurement 1; 6
measurement imprecisions... 13
measurement uncertainty 13; 30; 31
measuring geometry 25
metabolite 4; 61
method bias.. 21
method of additions 4
method performance 10; 22; 39
method recovery 30; 39; 40
method validation... 20
microextraction 54; 56
micro-extraction technique 48
minor elements... 10
monitoring recovery.. 7
motivation 8
multielement analyses 9; 12
multielement determinations 7; 13
multiresidue testing methods... 39

N

native analyte 8; 10; 15
native forms 11

native recovery 11
native residue 4
natural analyte 11
natural matrix 13
natural matrix Reference Materials 11
neutron activation analysis (NAA) 19; 24; 28; 45
neutron irradiated 24
neutron irradiation 24
nickel 26
NIST - National Institute of Standards and Technology USA 15; 18
non-statutory surveillance 62
normal distributions 2

O

occlusion 11
optimum of extraction time 49
organic residues 44
organochlorine pesticides 48; 56
outliers 16
outlying data 16
overall recovery 42; 59
overall uncertainty 13; 19; 31; 34; 35; 36
oxychlordane 48; 49; 51

P

palladium 26
partial recovery 42
peak identification algorithm 20
peak identification software 20
pentachloronitrobenzene (PCNB) 48; 49
performance characteristics 35
performance of the method 21
personal bias 32
PESTDATA 39
pesticide residue 1; 48
pesticide residue analysis 3
pesticides 4
physiological influences 6
pile-up 25
positive blank 20
precision 3; 18; 21; 30; 41; 59
precision error estimates 16
precision errors 14
precision estimate 18
precision requirements 40
pre-concentration 32
presampling 6
presampling factors 6
proficiency control materials 13; 15
proficiency materials 18
proficiency testing materials 11
propagation of error 17; 18
purification 42

Q

quality assurance 8

quality assurance program 5
quality control 5; 8; 11; 13; 22; 25; 27; 28; 38; 40; 41; 49
quantitative analysis 30

R

radio immunoassay..60
radioactive tracers 25; 26; 27
radioactivity..25
radiochemical neutron activation analysis - RNAA 24; 27; 28
radiochemical procedure...25
radiochemical separation...24
radiochemistry 24
radioisotopic tracer...24; 28
radiolabelled drug 44; 61
radiolabelled residue 44
radiotracers 24
random and systematic effects 30
random error 1; 3
random variations..31
range 19; 30
re-activation method...25
reagent blank...14; 21; 22
real blank 14
recoveries..4; 38; 60
recovery 1; 8; 9; 10; 13; 20; 21; 22; 27; 30; 39; 41; 42; 45; 59; 61
recovery calculation...2; 16
recovery corrected values 43
recovery correction 36; 44
recovery data 40; 60
recovery determination 12; 16; 25; 26
recovery estimate...43; 46
recovery experiments..42
recovery factor..........................1; 3; 4; 5; 7; 9; 11; 13; 15; 16; 18; 19; 22; 28; 30; 32; 33; 38; 43; 48; 49; 51; 54; 56
recovery factor uncertainties 19
recovery level 7
recovery materials 7
recovery measurement 12
recovery results 4
recovery solution 9
recovery study 20; 22
recovery testing 12
recovery tests...10
recovery) determination 25
recovery-corrected results...43
referee analyses..28
Reference Material..7; 11; 12; 13; 16; 21; 32; 33; 34
Reference Material uncertainties 19
reference method...28
reference technique 28
registration...38
registration of pesticides 38
regulatory laboratory...43
regulatory specifications 3
regulatory work...28
relative standard deviation (RSD)..3; 59
reliability 6
reliable analytical measurements 5
remedial action..22
repeat measurements 13
repeatability 33

repetitive determinations 16; 18
replicates 21
reported uncertainties 30
reporting limit 9; 12
reproducibility 27
residue 2; 11; 42
residue analyses 39
residue levels 38
residue limit 43
residue studies 38
result 30
result of a measurement 1; 31
retention 11
RM certification 28
routine analysis 5; 10
ruggedness testing 44

S

sample decomposition 13
sample extraction 42
sample manipulation 6; 7
sample matrice 31
sample matrices 33
sample preparation 42
sample pretreatment 42
sample storage 7
sample weight test 21
sampling 6; 7
sampling model 6
screening test 43
selenium 26
separate uncertainties 35
separation of analyte 6
sewage 28
significance test 34; 35
siliceous residue 11
silver 26
single determinations 2
single value 30
sludge 28
soil dissipation studies 40
solution addition 9
sources of error 6
sources of uncertainty 31
speciation studies 24
specificity 24
spike 9; 10; 15; 17; 33; 49
spike addition 17
spike level 49
spike recovery 16; 18
spike recovery studies 11
spiked control materials 18
spiked analyte 10; 14; 15
spiked material 1
spiked recovery sample 10
spiked sample 7; 10; 12; 15; 46; 60
spiking 8; 9; 12; 13; 20; 22; 30; 60
spiking levels 9; 10
spiking studies 34
stability data 39

standard deviation 1; 2; 3; 4; 16; 20; 31; 34; 35; 41; 59
standard method 33
standard uncertainty 31
standardized practice 22
statistical calculations 16
statistical control 8
statistical outlier 39
statistical treatment 6
stripping voltammetry 21
Student's t 34
sulfadimidine (sulfamethazine) 43
systematic effect 35
systematic error 1; 5; 16; 19; 22; 31

T

technical competence 12
thallium 26
thorium 27
tin 27
tolerance level 39
tolerances 3
total element concentration 5; 11
total recovery 4
total residue studies 44
trace analysis 28
trace element 24
trace element analysis 10
trace levels 5
traceability 24
traceable 12
tracer method 25
trained personnel 8
true value 3; 19
trueness 30
tungsten 27

U

U.S.Environmental Protection Agency (EPA) 49
unacceptable error 21
unadjusted 22
uncertainties 5; 18; 33
uncertainty 10; 13; 16; 18; 22; 24; 27; 30; 35
uncertainty calculations 16
uncertainty components 24
uncertainty computations 18
uncertainty estimate 18; 36
uncertainty estimation 30
uncertainty in the reference value 19
uncertainty of the recovery factor 19
uncertainty on recovery 35
uncorrected 22
uncorrected analytical results 43
uncorrected background 21
uncorrected individual value 3
uncorrected residue values 39
uncorrected result 1
United States Environmental Protection Agency (US EPA) 38
uranium 27

V

valid analytical data...8
validated methods 40
validation of a method 59
validity...6
value 1
values observed 34
variability...3
veterinary drug residue...3; 42
veterinary drug residue analysis..59
veterinary drugs...4; 43
violative residues 43
volatile elements..25
volatilization losses 6

W

worst case assumption 44

X

X-ray emission spectrometry..19

Z

zearalane..45
zeranol..45
zirconium 27